Railway World

YEAR BOOK

Previous page:
This must be the oddest set of points in the British Isles! But where is it? See the article which begins on page 13 to find out. *Colin Boocock*

This page:
The morning sun highlights the glossy surface of preserved 4-6-0 No 5305 as it approaches Banavie on 2 September 1989 with a West Highland line enthusiasts' special. How is such an appearance achieved? See the feature article which begins on page 26.
Dr W. A. Sharman

1991

Railway World YEAR BOOK

Edited by Colin Boocock

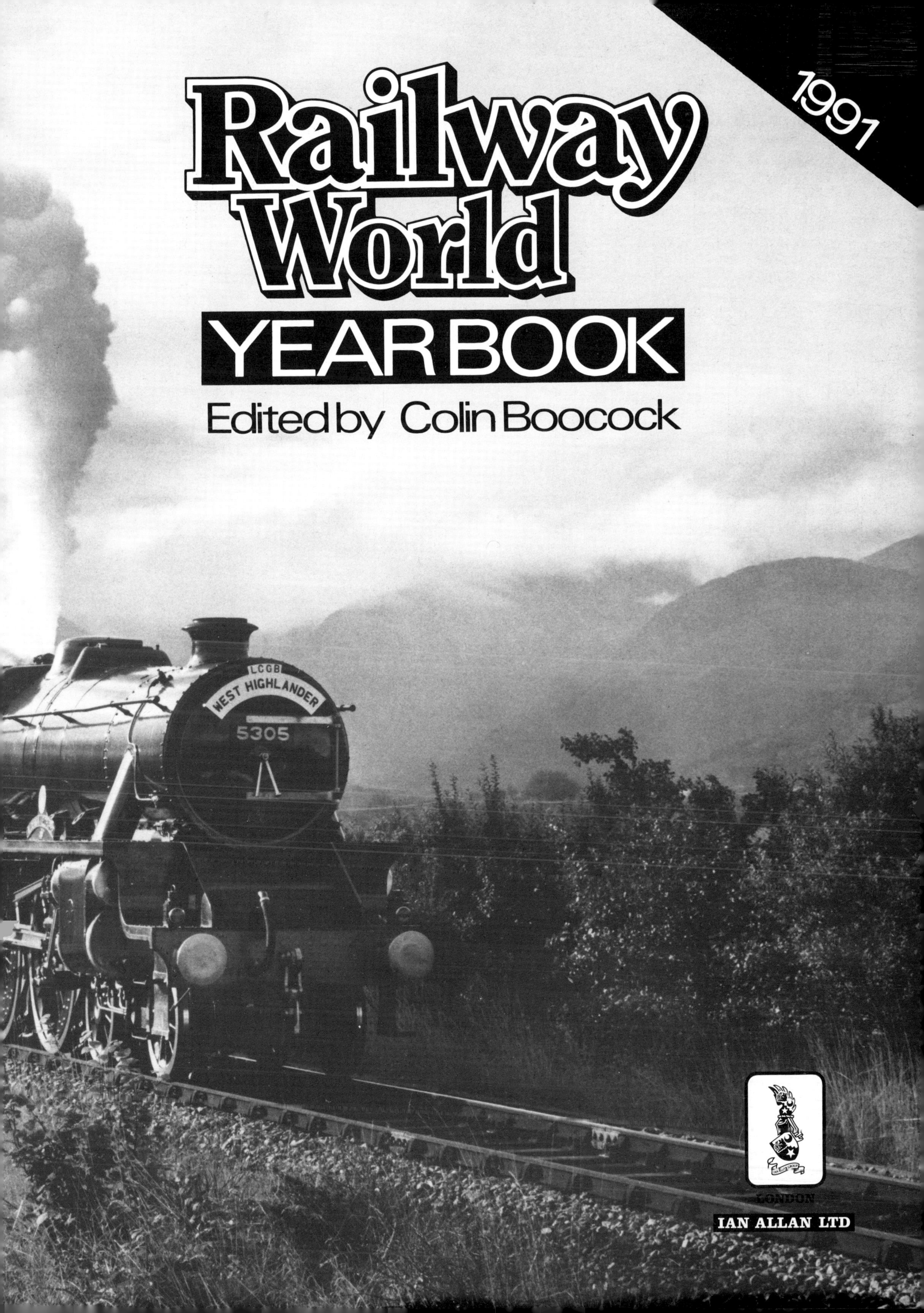

LONDON
IAN ALLAN LTD

Contents

Front cover:
LNER 'K1' No 3442 *The Great Marquess* awaits departure from Appleby in March 1990. *John S. Whiteley*

Below:
Even with the quantity of steam locomotives to be seen on British Railways in the 1950s it was still possible for unusual workings to occur. Otherwise, why would the 5.24pm stopping train from Eastleigh to Southampton Terminus be graced by Churchward 2-8-0 No 2821 on 7 May 1959? *Colin Boocock*

First published 1990

ISBN 0 7110 1932 0

Published by Ian Allan Ltd, Shepperton, Surrey; and printed by Ian Allan Printing Ltd at their works at Coombelands in Runnymede, England

Railtalk

Colin Boocock

This is the first *Railway World Year Book* 'Railtalk' section to have been written at 125mph in a British Rail Mk 4 coach!

1991 is the year of completion of East Coast main line electrification. This 1991 issue supports that event with an article about the new Class 91 locomotives and Mk 4 rolling stock. Have *you* ridden the InterCity 225 yet?

Quite possibly you also enjoy riding steam-hauled trains, either preserved, restored or even the miniature variety. Dip into this book and revive your recent or distant memories.

Or perhaps you have a particular liking for the unusual locomotives on our railways, from compounds to odd and faintly idiotic tank engines. There is much to whet your appetite in this book.

Did you know that BR pumps nearly 5,000 gallons of water per minute out of the Mersey Railway tunnel? Where does it go? Is it true that somebody actually *buys* it?

What was the first main line to be built a cross the European Alps? What are the railways like in New Zealand?

These are not questions in our popular railway quiz (see page 76). To answer these questions you might think you need to be a highly intelligent, well travelled railway historian (which you may well be). But all you actually need do is read this *Year Book*. There is something in it specially for you.

Acknowledgements

Once again I have been showered with riches by contributors to *Railway World Year Book*. I thank particularly those authors and photographers who supplied material to my invitation.

A special 'Thank you' goes to Alan Thorpe for another selection of cartoons. These are made available by courtesy of Tony Jervis, Editor of *BRC News*, the journal of the Bournemouth Railway Club.

Right:
Electric working of East Coast main line express trains began in 1989. Class 91 Bo-Bo No 91004 rests at King's Cross having propelled the 18.10 InterCity train from Leeds. *Colin Boocock*

The Mersey Railway Tunnel

Delwyn Bryan

The Mersey Railway lays claim to two historical records. It was the first under-river railway in the world, and it was the first to be converted from steam to electric traction.

During the 18th century Liverpool was linked with the Wirral peninsula only by means of ferries propelled by sails. In the early 19th century paddle steamers were introduced, but the rapid industrial growth on both sides of the river quickly outpaced this form of cross-river transport. The problem of overcrowding was further aggravated in bad weather.

The idea of linking Liverpool with Birkenhead by means of an under-river tunnel had been conceived as early as 1825 but, due to scepticism, this proposal did not materialise. The first serious plans were drawn up in 1864 when the Liverpool & Birkenhead Railway Co petitioned Parliament for powers to construct a tunnel under the River Mersey. There was fierce opposition from the established railway companies in both Liverpool and Birkenhead at that time, whose existing railway routes would be seriously affected by the tunnel, and the Bill was defeated in Parliament in 1865.

The Mersey Railway Co was formed in 1870 and the following year the Mersey Railway Act was granted. Work was immediately started on the Birkenhead side by sinking a shaft to a depth of 90ft to determine the underlying rock structure. Unfortunately, as history so often relates of these projects, the company engaged to carry out the engineering work went into financial difficulties and the work came to an abrupt halt.

In 1879 the Mersey Railway Co appointed Major Isaacs to undertake and finance the complete project. Work restarted in December of that year. The contract for excavating the tunnel was sublet to John Waddell & Sons of Edinburgh who appointed Charles Fox and James Brunlees as engineers.

Initially, two pump shafts were sunk, one on the site of the old

Left:
Not all towers built by the Victorians were decorative. This one at Hamilton Square station, Birkenhead, was built 120ft high to locate a water tank with sufficient head of water to drive the hydraulic lifts which connected the underground Mersey Railway platforms with the surface. *BR*

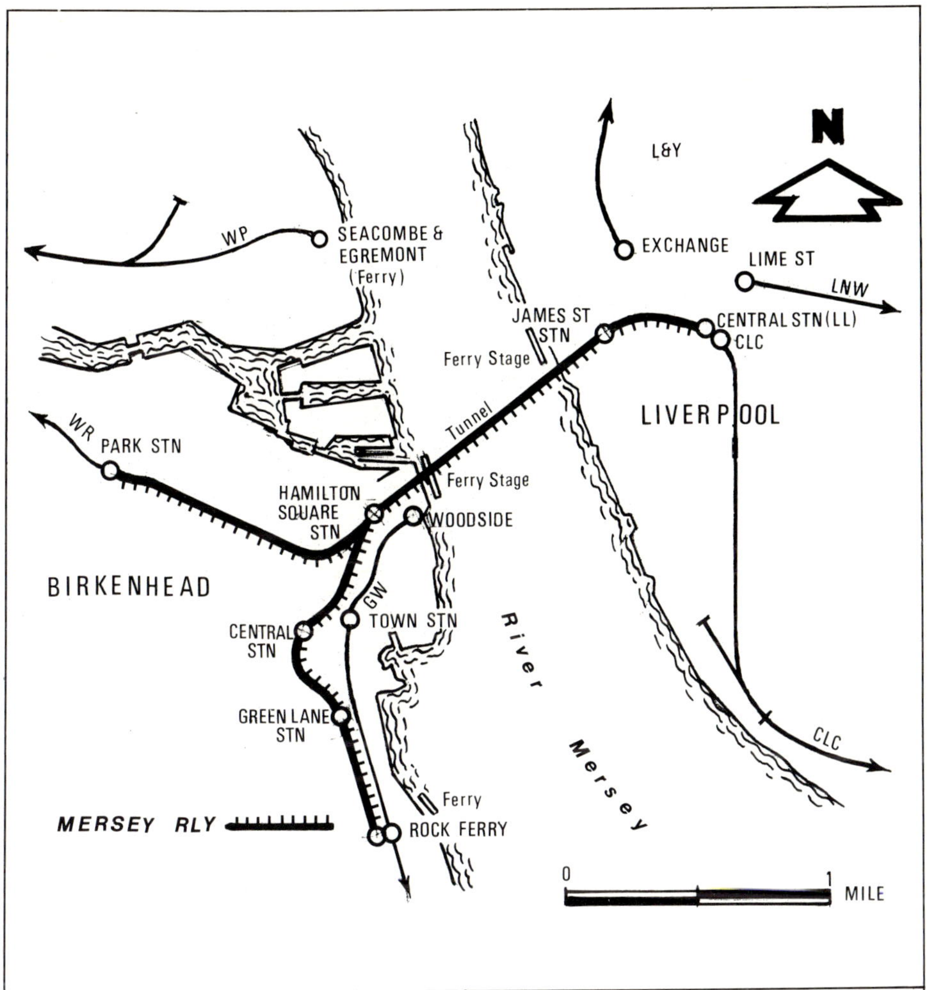

Left:
Copy of an early sketch plan taken from a drawing of the Mersey Tunnel dated 1906.

George's Dock in Liverpool and the other at Shore Road, Birkenhead. Each was sunk to a depth of 170ft, the one at Liverpool having a cross-sectional diameter of 15ft, and that at Shore Road a diameter of 17ft 6in. Large drainage pumps were located in these shafts.

From the bottom of each shaft a 7ft drainage tunnel, or 'heading', was driven at a slightly ascending gradient towards the middle of the river, prior to the construction of the main tunnel. The construction work involved the boring of three tunnels in all: the main railway tunnel, the previously mentioned drainage tunnel and a ventilation tunnel. Both the ventilation and the drainage tunnels merge into one another under the river bed section, approximately 8ft below the main tunnel and to one side.

In 1883 a compressed-air boring machine, invented by Col Beaumont, was introduced into the workings on the Birkenhead side. This machine was capable of cutting a 7ft diameter of tunnel at a rate of 17yd a week. The rifle marks formed by the cutting tool of the machine are still visible on the wall of the ventilation heading.

The main railway tunnel was lined with eight rings of brickwork, for which purpose a total of 38 million bricks were required. The height of the tunnel was 19ft, and the width was 26ft in order to accommodate the two standard gauge tracks. The trackbed consisted of small pieces of sandstone rock with a layer of cinders on which was laid the sleepers. The tunnel descends from James Street at a gradient of 1 in 27, then levels out under the river bed, and rises at 1 in 33 up to Hamilton Square station. At the mid-river section the crown of the tunnel is only 33ft below the river bed.

On 17 January 1884, the two headings met under the river at a distance of 1,115yd from the Birkenhead shaft. By the end of 1885 all the associated work involved in the tunnelling and the laying of the track had been completed.

The Board of Trade's formal inspection was carried out by General Hutchinson on 4 January 1886. He remarked on the achievement and commended both the engineers and the contractors. The tunnel was opened on 24 January 1886, the ceremony being performed by the then Prince of Wales at James Street station.

Passenger services began on Monday 1 February 1886, running between Green Lane station on the Birkenhead side and James Street station in Liverpool. Two years later the Birkenhead Park branch was opened and in 1891 the line from Green Lane was extended to Rock Ferry. The final section of the Mersey Railway from James Street to Liverpool Central was completed in 1892.

Simultaneously with the construction of the Mersey Railway, there were other railway developments on the Wirral. The Hoylake Railway planned a railway to West Kirby. Work on an extension to New Brighton began in 1882 and the following year the Wirral Railway was incorporated with the object of building a line to link Birkenhead North with the Mersey Railway at Birkenhead Park. This link with the Mersey Railway came to fruition in January 1888.

The Mersey Railway carried 10 million passengers in 1890 but after that year the numbers began to fall. It was not difficult to find the reasons for the decline. Due to severe gradients in the tunnel, powerful steam locomotives were necessary to haul the trains, which resulted in both the totally enclosed underground stations at James Street and Hamilton Square being filled continuously with a choking mixture of smoke and steam. The ventilation system installed at both these stations, at great expense, failed to overcome the problem and in 1899 the Mersey Railway proposed to electrify the line.

Electrification was completed in 1903 when the Mersey Railway became the first steam railway to be converted from steam to electric traction. It was a notable feat that all the conversion work had been undertaken without any disruption to the normal railway services. The electrification was the 650V dc fourth rail system and was carried out by the British Westinghouse Electric & Manufacturing Co of Manchester. The new power house, generating units, steam-raising plant and all associated equipment were located at Shore Road adjacent

Right:
The original pair of Andrew Barclay beam pumping engines at Shore Road, Birkenhead.

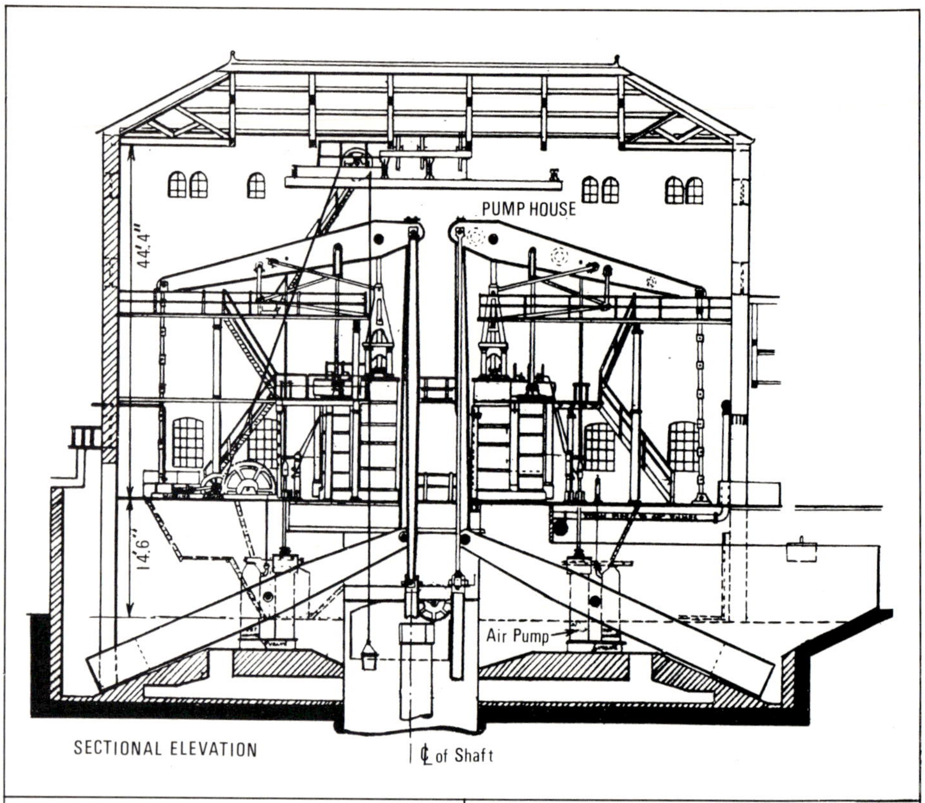

to the pumping station in Birkenhead.

On the Liverpool side of the Mersey, the Lancashire & Yorkshire Railway lines between Liverpool Exchange, Southport and Crossens were electrified in 1904 and the section between Liverpool, Aintree and Ormskirk was converted in 1913.

It was not until 1938 that the London, Midland & Scottish Railway (LMS) was to electrify its routes on the Wirral to New Brighton and West Kirby at 650V dc.

Because of the depth below street level of the stations at James Street and Hamilton Square, passenger access to the railway was by means of six hydraulic lifts, three at each location. These lifts were installed by Easton & Anderson and were operated by large direct-acting hydraulic rams which either pushed the lift car up to street level or gently lowered it down to the bottom landing.

Each lift had a passenger capacity of 100 people and measured 20ft long, 17ft wide and 9ft high. There were seats around the inside walls. The maximum speed of the lifts was 120ft/min and the time taken to travel from the bottom hall to the top hall, a distance of 80ft, was 45sec. The lifts were controlled by an attendant in each car who operated the control ropes and the access gates.

The large water tanks which were necessary to provide the essential head of water for hydraulic lift operation were located 120ft above street level on two large brick towers at James Street and Hamilton Square. These tanks were replenished by steam-operated ram type pumps. There were three pumps at each location, each capable of pumping 500gal/min. That rate of pumping is insignificant, however, when compared to what the tunnel drainage pumps could do.

Extraction of water from the tunnel system has always been one of the major operations with which the Mersey Railway (and now Merseyrail) has had to contend right from the start of the construction of the tunnel in 1879. Two pumping stations were constructed to cater for the inflow of water, located at George's Dock and Shore Road.

Initially the water was extracted by six large steam-operated pumping engines, three at each location. At George's Dock the pumping plant consisted of two Hawthorne Davy pumps and one of Andrew Barclay manufacture. The largest of these pumps could shift 3,500gal/min.

At Shore Road there was one 30in Hawthorne Davy pump and two Andrew Barclay pumps similar to those at George's Dock. These were affectionately known as a 'grasshopper' type engine due to the bobbing movements of the main beam at the connecting rod end which simulated the jumping action of a grasshopper. These engines were most impressive machines each weighing a total of 262 tons and towering 50ft above the base. Although each beam engine was at ground level, the 40in diameter plunger type pump which it operated was located 170ft below at the bottom of the shaft. The reciprocating action to the plunger was by means of a long wooden connecting

Left:
The electrification of the Mersey Railway brought some quite splendid coaching stock to the line. This first class driving trailer vehicle poses outside the manufacturer's works of Cravens of Sheffield. It is one of a number of replacement vehicles purchased to replace those original vehicles damaged in World War 1. *BR*

rod the full length of the shaft. The tendency for the long rod to flex was kept in check by wooden guides located at regular intervals down the shaft.

In operation the pumps were practically silent, apart from the low strumming action of the long wooden rods when they took the pumping load. The speed of a pump was controlled by the main steam regulator valve, whilst the length of the stroke could be varied by adjusting the tappet wheel which operated the steam valves through a system of levers. After completion of a stroke, the period for reversal of the pump could be varied by adjustment to the oil dash-pot unit which determined the delay between the operation of the admission and exhaust valves.

The main beam was pivoted on a live fulcrum and constrained by 'Watt' parallel motion links, thus maintaining a truly vertical motion to the piston rod and the wooden connecting rod to the pump. The engine was double acting and it was therefore necessary to counterbalance the heavy working parts. This was achieved by a balance beam which was located below the main working level, pivoted at the centre, with one end connected via a system of links to the main beam, the other end being loaded with 20 tons of steel ingots.

An engine of this type survived in standby duty right up to 1959.

From the late 1920s steam pumping slowly gave way to electric pumping. The present day pumping station at Shore Road consists of two electrically driven borehole pumps, each with a pumping capacity of 4,200gal of water per minute. The pumps were provided by Sulzer Bros and are driven by 310hp motors and operate off the 650V dc traction supply. The pumps start and stop automatically at predetermined water levels in the well, operated by float switches.

Similarly at George's Dock there are two dual duty borehole pumps (3,500 and 5,500gal/min) which work on 415V supply direct from the local Electricity Board. Due to the influx of water into the tunnel one 3,500gal/min pump normally runs continuously, and the second is on standby duty. Under normal conditions the water is sold to the *Liverpool Daily Post*! It is pumped via a 16in diameter PVC main pipe up to the newspaper's works where it is used for cooling the presses and is then finally discharged into the town sewers. There is also a facility, by operating a bypass valve within the pump house, to divert the full pumped output into the adjacent Canning Dock.

While it is not the purpose of this article to deal in depth with the rolling stock of the Mersey Railway, it is pertinent to note some of the more unusual features since electrification. The early Mersey Railway coaches were clearly influenced by American design, having features such as clerestory roofs, open end verandahs (later closed in), slab sides and centre couplers. A real oddity for the UK was that air for the air brakes was fed to the trains at the terminal stations from shore-based compressors, the air in the carriage reservoirs having to suffice for braking at all the intermediate stops, including any signal checks!

The Wirral lines were electrified in 1938, and through working was set up between the new Wirral sets and the Mersey Railway units. In 1955 the whole route was converted to third rail working in readiness for the new electric multiple-units (EMUs) which were to replace the old Mersey Railway sets in 1956.

The present situation with regard to traction electricity supply is that the 650V dc current is fed to the third rail from lineside substations which are located approximately 2 miles apart. The primary voltage fed to these substations is three-phase 11,000V ac which is transformed and rectified to 650V dc and fed through high speed dc circuit breakers.

The 11,000V ac supply is obtained from the local Area Board at Shore Road. There is also a feed on the Southport line at Bank Hall, and in the event of failure of one feeding point, circuits can be linked to the other to maintain continuity of traction supply.

Control of the electric supply is now concentrated at one control room, at Hall Road on the Southport line. This uses modern visual display equipment.

Below left:
This is Shore Road pumping station at Birkenhead showing the renovation work in progress and the new domed public access to the museum part of the pumphouse. The remaining beam engine has been restored. *BR*

Below right:
The two modern Sulzer pump motors at George's Dock, Liverpool, drive pumps which are located 165ft below ground level. The drive is via a long stainless steel shaft of 81mm diameter. Potential flexing of the long shaft is kept in check by means of special bearings at regular 10ft intervals within the rising main. *BR*

Right:
Following electrification of the LMS suburban lines on the Wirral, through working became the norm between the Wirral and Mersey routes. This train of Wirral stock was photographed at Green Lane on the former Mersey Railway while working from Rock Ferry to Liverpool Central on 20 June 1984.
Colin Boocock

Maintenance of the Mersey Railway stock was formerly carried out in the cramped depot at Birkenhead Central, but in 1938 a new depot was opened at Birkenhead North which was capable of carrying out heavy lifting and important component repairs, as well as routine maintenance. The depot has recently had its role extended to the full overhaul of the whole fleet of EMUs working on the Wirral and other Liverpool routes.

Major changes were to be made to the Mersey railway network in the early 1970s with the proposals by the Merseyside Passenger Transport Authority to rearrange the railways in Liverpool and Birkenhead to provide through running where possible, and better connections where not. Thus was born the Link line which joined up the Northern Line routes from Garston in the south to Southport in the north, on the Liverpool side of the river; and the Loop line which was formed by extending the Mersey Railway Tunnel inland and round a largely circular route to link up underground with Liverpool Lime Street and Central stations, crossing under the Link line at a new station at Moorfields and under Central station. There was also a new tunnel junction at Hamilton Square (known locally as the 'dive under') to separate trains taking the Birkenhead Park route from those arriving from Rock Ferry.

The Loop line is a single bore tunnel, 2 miles in length and 15ft 5in in diameter. It runs in a clockwise direction from James Street's new Loop platform through Moorfields' deep level platform, round to Lime Street deep level platform, and Central station underground before rejoining the old Mersey Railway Tunnel under the Derby Square section of the city, and finally into the original Mersey Railway down platform at James Street. The tunnel varies in depth between 58ft and 124ft below the city and is lined with concrete.

A complete new station was constructed at Moorfields to replace the old Lancashire & Yorkshire Railway terminus station at Liverpool Exchange. The movement of passengers between the Loop and Link platforms and the street levels is by means of escalators. There are also lifts of 10 persons capacity which link all levels for the benefit of disabled passengers.

The new Loop station at Lime Street has its concourse under St

Below:
Plan and elevation diagrams of the Mersey Railway Tunnel showing drainage and ventilation headings.

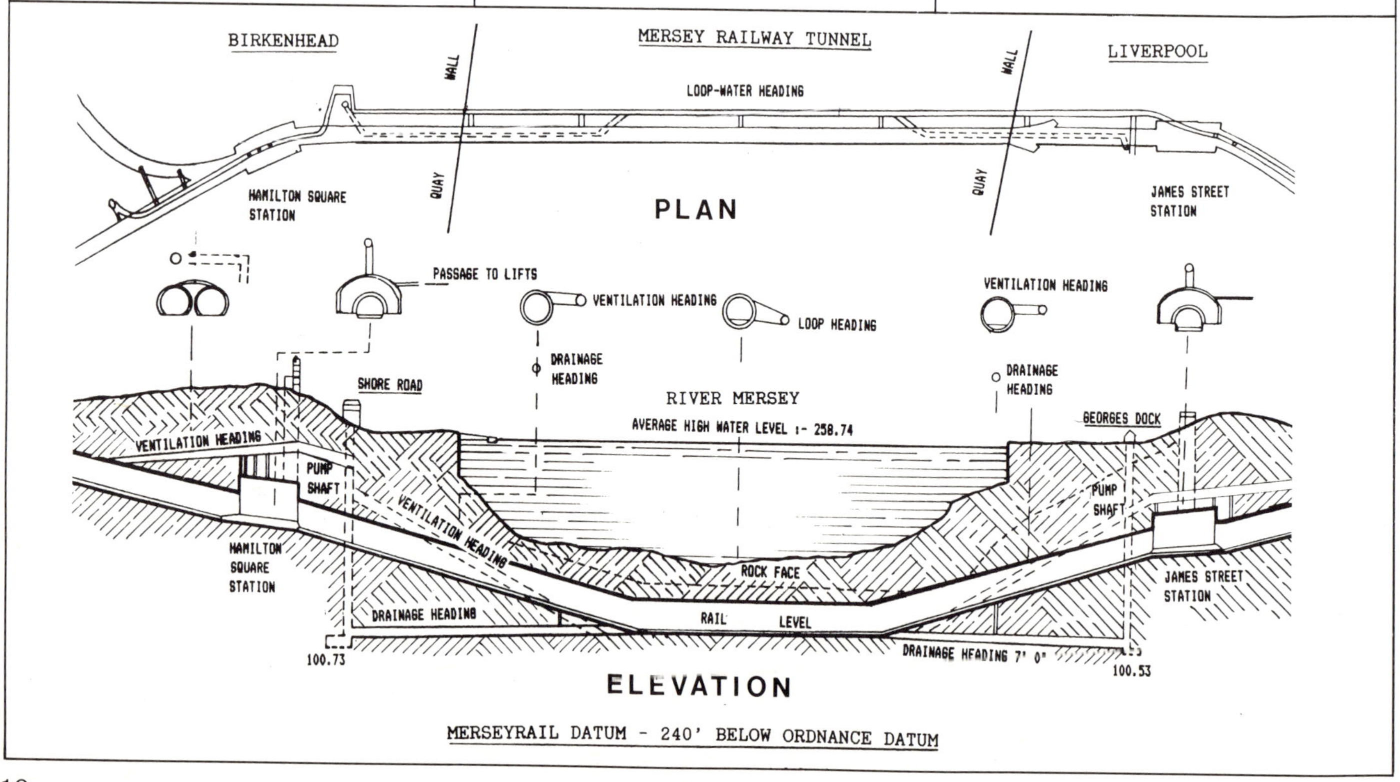

George's plateau. Customers gain access to the deep level platforms by escalators, and there is a further bank of escalators linking the concourse with that of the main line station above ground. Liverpool Central station has two platform levels. The low level platforms (nonetheless higher than the deep level) which cater for the Northern Line trains utilise the former Mersey Railway terminal platforms, which had been released by the diversion of the Wirral Line through the Loop. The Loop platform at Liverpool Central is at deep level.

Both James Street and Hamilton Square stations were modernised and were provided with additional platforms. The old up platform at James Street has been withdrawn from normal passenger service and is used only for emergency purposes, for example turning back trains when the Loop tunnel is closed for any reason. The former Mersey Railway Tunnel between James Street and Liverpool Central is still intact and forms the connection between the Loop and Link for exchanging rolling stock between the Northern and Wirral Lines and for maintenance or overhaul at Birkenhead North depot.

The concourses, subways and platform areas of the underground stations are all equipped with closed circuit television for anti-vandalism and crowd control surveillance. The control room at Moorfields monitors the 24 TV screens 24hr a day, as well as making announcements as necessary on the public address systems.

The underground railway system is controlled through a central power signalbox located at James Street station. Signalling is three-aspect colour light on the Link line, while the Loop line has two-aspect signalling.

With the construction of the Loop and Link lines a further five pumping stations were introduced into the system. These are of minor capacity in comparison with the main ones at Shore Road and George's Dock, and are located at the 'dive under' section, Mann Island, Central station (Loop) and (Link), and Leeds Street bank.

During the construction of the Mersey Railway Tunnel the inflow of water was in the region of 9,000 gal/min. By the late 1920s this had fallen to 5,000gal/min. In the 1970s the water flow had further reduced to 4,300gal/min due to fissures in the tunnel wall silting up and also to the decline in river shipping traffic reducing disturbance to the river bed. However, the present-day combined output from the George's Dock and Shore Road pumps has increased again to 4,700gal/min. This has been brought about by the latter-day decline in industry on Merseyside leading to lesser use of water, and the consequential rise in the water table.

Above:
Built before World War 2, the rolling stock repair shop at Birkenhead North provides excellent facilities for the repair and overhaul of electric multiple-units on Merseyside. Virtually all overhauls to the modern fleet of Classes 507 and 508 units are carried out there now. *Colin Boocock*

In 1985 it became essential to implement a 'track de-watering scheme' to supplement the track drains in the Loop tunnel system. The water level in some sections had risen so much that track circuits were affected and signalling failures reached an impossible level for reliable train operation.

The de-watering scheme involved the installation of eight small electric submersible pumps in the inspection manholes of the track drain in the section of tunnel between James Street and Moorfields. A further six were put in place between Liverpool Lime Street and Central stations. The capacity of these small pumps is 60gal/min and all discharge into two 8in gravity pipes attached to the tunnel wall. The water is discharged into the ventilation heading and into the main sump of the Central Loop pumping station. In 1986 it was necessary further to extend this system, and two more pumps were inserted beyond Moorfields station. The water table is expected to continue to rise for a few more years until it reaches a level similar to that in pre-industrial times.

Just as the early Mersey Railway electric trains were replaced during BR days by the new EMUs of 1956 vintage, so have the 1938 Wirral and 1956 EMUs in turn been displaced by modern BR EMUs of Classes 507 and 508. Normally the Wirral line, which incorporates the tunnel under the Mersey, uses the majority of the Class 508 sets.

Recent developments include the electrification beyond Rock Ferry to Hooton, with future aspirations to reach Chester and Ellesmere Port, even though these places are outside the Merseyside PTE area. There is currently some excitement over the PTE's agreement to support upgrading of the signalling and control arrangements. There will be a new combined control centre covering centralised traffic control of all signalling and of traction electricity supply, elimination of Victorian semaphore signalling on the outdoor sections, and modernisation of level crossings. The new control centre will also embrace passenger information and surveillance equipment, and will cover both the Northern and Wirral line services.

Thus the Mersey Railway Tunnel now forms a central part of a network of railway routes serving Liverpool and Birkenhead and beyond in a professionally integrated, modern and highly successful way. The dreams of its promoters, over 110 years ago, must surely have been realised in full measure.

Odd Engines 1:

The Crosti '9Fs'

British Railways built 10 Crosti 2-10-0s to test the potential for improved thermal efficiency of the '9F' 2-10-0 design. It is however, difficult to improve on the excellent, and the Crosti '9Fs' were no exception to this rule.

Signor Crosti's idea was to preheat the boiler feed water by extracting yet more heat from the exhaust, and not releasing so much through the chimney. The preheater drum under the boiler barrel enabled the exhaust gases to be drawn through tubes surrounded by feed water. The multiple exhaust chimney (if that was the correct term) was in front of the cab at the right-hand side.

Being heavy freight engines they often ran cool, and the drums corroded badly through acid formation. The locomotives were soon converted to normal practice by removing the drums and reverting to the standard exhaust arrangement. In this form, with their slightly smaller boilers and oddly-shaped smokeboxes giving away their background, the rebuilt 'Crostis' ran successfully.

Above:
Crosti 2-10-0 No 92020 is stabled at Cricklewood on 17 November 1957. The front of the preheating drum can be seen below the smokebox door, the water feed pipe to the boiler barrel mounted clack valve is lagged, and the 'chimney' on top of the smokebox is a dummy (or was it used for lighting-up purposes?).
Colin Boocock

Centre right:
The exhaust system of a Crosti '9F' 2-10-0 was on the right side of the locomotive in front of the firebox. This photograph shows the multiple jets needed to get enough draught to pull the smokebox gases through both the boiler and the preheater drum. Below the exhaust pipe is a water drain to take away condensate from the pipe feeding the exhaust steam injector. The smoke deflector was added to draw away steam which drifted into the fireman's cab front window. *Colin Boocock*

Bottom right:
The restoration of the Crosti 2-10-0s to 'normal' condition was done with relatively little expense. The smaller diameter boiler was retained, apparently with no ill-effects. Marginal cosmetic changes were made to the valances and front plating, to fill in holes. Otherwise the locomotives were still ugly, as seen in this view of No 92021 passing Kegworth on an up freight on the Midland main line on 5 September 1960. *Colin Boocock*

Vintage Rail Travel in the Isle of Man

P. W. B. Semmens

Above:
The 1960s was the last decade of virtually full operation of the Isle of Man Railway. Rust-red-painted Beyer Peacock 2-4-0T No 10 *G. H. Wood* leaves St John's with the Ramsey portion of a train from Douglas to Peel and Ramsey on 31 August 1962. The Peel branch is the track in the foreground. *Colin Boocock*

A century ago, when Britain led the industrialised world, the rising numbers of the middle-classes that owned or managed our factories began to have both the money and opportunity to take holidays. Many of the country's wealth-producing centres were situated along the fringes of the Irish Sea, and for those who wanted the excitement — and prestige — of an overseas holiday, but who were worried about the language problems on the Continent, the Isle of Man had much to offer. People could travel there by steamer from several mainland ports, and, once on the island, there were transport systems which were well adapted to cater for the tourist business of those days. Some, indeed, were specifically constructed for just that purpose, and the survivors are still well placed to offer today's more mobile holidaymaker the opportunity to experience what rail travel was like in Victorian and Edwardian times.

Although the Isle of Man is only some 30 miles long by 11 miles wide, it is remarkably hilly. The highest peak is Snaefell, rising 2,036ft above sea level, and is located only four miles from the coast. It is these central hills that made transport *across* the island quite difficult, as well as restricting communications along parts of both the eastern and western seaboards. Travel by sea had thus been important in the early days, with many sheltered and well-fortified ports developing all round the island.

With the coming of the railway age, it was natural that thought should be given to improving overland transport, but it was not until 1873 that the first line was opened, running for 11½ miles across the middle of the island from Douglas to Peel. In a hilly island the advantages of a gauge less than the standard 4ft 8½in were obvious, and a distance of 3ft between the rails was chosen. Just over a year later, a second radial route from Douglas was opened, this time running for 15½ miles to Port Erin via the original capital of Castletown. The need for better transport in Douglas itself was acknowledged when their horse trams were inaugurated along the promenade in 1876. They also are still with us, but the cable car system that ran inland from the harbour has long disappeared, although there are a few relics of it still on display at the horse-tram carsheds.

These developments left the large northern port of Ramsey without a rail connection, but there was no way in which a direct steam-operated line could have been built between it and Douglas, although one was actually authorised. However, in 1879 the 16½-mile Manx Northern Railway opened from St John's inland from Peel, and followed the west coast to Kirk Michael, before swinging eastwards round the northern edge of the hills to reach Ramsey. Out of the junction the two lines ran parallel for the first half-mile, enabling the two portions of combined double-headed trains from Douglas to indulge in some competitive running over this stretch after they had split.

At one time there were extensive mining activities on the Isle of Man, and the final section of the main steam-worked system was opened in 1886 from St John's up the steep valley to the mines at Foxdale. While all the other steam locomotives on the island were 2-4-0Ts, which we will meet later in our story, the gradients on this branch required something larger. An 0-6-0T *Caledonia* was thus supplied by Dübs to deal with the freight traffic on this part of the system, and became a well known sight blasting its way over the flyover and up the bank.

The next railway system to be constructed on the island not only introduced a new, state-of-the-art technology, but was designed for two roles that do not sound out of place today. One was to provide a tourist attraction, and the other was to increase land values as part of the general development of the clifftop area at the north end of Douglas Bay. In 1892, only 13 years after Werner von Siemens had demonstrated his rudimentary electric train at a Berlin exhibition, the Manx parliament granted permission for the building of Howstrake estate. The development company's authority included the right to construct a tramway, and this continued beyond the northern edge of the new estate as far as Groudle Glen, one of those wooded valleys leading down to the sea for which the east coast of the island is noted. When, however, it opened in 1893, the cars were found to have blazoned along their sides the imposing title 'Douglas & Laxey Coast Electric Railway', and the line was extended over the further four-mile stretch to the village of Laxey within the next year.

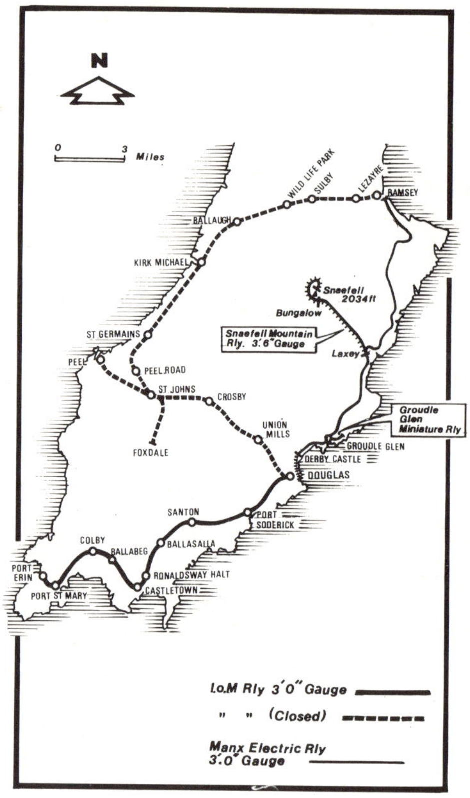

Above:
Map of railways on the Isle of Man.

The original electric tramcars, with their twin 25hp dc electric motors, were capable of hauling a trailer over the 1 in 25 gradients of the line, and immediately proved to be extremely popular. It used to be said that holidaymakers, having booked their tickets for the tramway at the Derby Castle terminus, used to take the horse trams half a mile back along the promenade to join the end of the queue to get on the next available service!

Some of the people connected with the tramway then developed an even more ambitious scheme for an electric railway from Laxey to the summit of Snaefell, which involved a ruling gradient of 1 in 12 — more than twice as steep as that of the tramway.

Although they were confident that the electric motors could take the cars up the mountain relying on adhesion, some more effective form of braking was required for the descent. A raised centre rail of the Fell type was accordingly provided, with slipper brake shoes which required changing every three or four trips. To provide space to motorise the Fell rollers should this turn out to be necessary, the gauge of the track was widened to 3ft 6in. This prevented the Snaefell cars running over the coastal line, although nowadays they are mounted on temporary 3ft gauge bogies whenever they need to be taken to the main depot at Douglas for overhaul.

The Snaefell Mountain Railway also proved to be a popular tourist attraction, but there has only been enough demand to justify operations during the summer. Indeed, on the top stretch of the line, the overhead contact wires have been taken down for protection every winter except one, the deliberate omission of the one year conclusively proving that need for the operation! Meanwhile the original tramway company was pressing on with the extension of the coastal line to Ramsey, which was finally reached in 1899. This was only achieved by stretching finances to the limit, and the withdrawal of mainland support for a Manx bank forced the Isle of Man Tramway & Electric Power Co into liquidation early the following year. However, the line was taken over by a London-based company in 1902, and, against all the odds, the whole of the route is still in operation, as the Manx Electric Railway, nearly 90 years later.

After the electric tramway had reached Groudle Glen, another entrepreneur extensively developed its natural beauties as a tourist attraction. In the late Victorian era, holidays were far more genteel than those we now enjoy. A quiet walk through the leafy glades of a wooded valley alongside a swiftly running stream was considered highly relaxing and delightful, with the possibility too of getting refreshment at the entrance hotel or in the cafe further along. Clearly the arrival of the latest in the way of electric transport would bring a continuous stream of people right to the doorstep, but how do you persuade them to stay, rather than taken an early train back? The answer at Groudle Glen was to build a zoo on the cliffs above the sea, where sea lions, bears and birds were kept. For those who might feel that the walk round the headland was too long or exposed, a 2ft gauge railway was built to carry passengers. A diminutive, Bagnall-built 2-4-0T locomotive named *Sea Lion* arrived in 1896, and was joined by its sister, *Polar Bear*, nine years later.

The complex rail transport system of the island served the holidaymakers and residents well for the first half of the 20th century. The Foxdale branch closed in World War 2, but the introduction of holidays with pay saw the popularity of the Isle of Man as a holiday resort reach a new peak in the 1950s. However, the technological and social changes of the late 1950s and 1960s were to produce a drastic switch as visitors turned to their own cars, or flew to Mediterranean resorts for their holidays.

The first of the systems to feel the pinch was the Manx Electric Railway, which was sold to the island's government in 1957 at a rock-bottom price, but this only postponed the financial problems for a couple of decades. In 1976 just the Douglas-Laxey and Snaefell lines were worked, and during the summer only. The results showed that the closing of the continuation to Ramsey was a false economy, and it was reopened for the 1977 season. 1979 marked the centenary of Siemens' electric line in Berlin, and the Manx Electric Railway was able to capitalise on this, the resulting interest generated in its historic system enabling it to refurbish many of the vehicles to reduce operating costs without altering their vintage appearance. Year-round operations have resumed, but staff flexibility has enabled costs to be kept down, with people from the workshops taking traffic duties on the line in the summer.

With these difficulties arising on the tramway, the writing was also on the wall for the island's small and

Left:
The Edwardian tramcars of the Manx Electric Railway are an attraction to visitors, but some modernisation has in fact taken place. Note the vertical hydraulic dampers on the outside of the bogies on car No 5, photographed at Laxey on a working to Douglas in the summer of 1989. *P. W. B. Semmens*

roundabout railway system, whose steam locomotives had been joined in 1959 by the two articulated railcars made redundant by the closure of the County Donegal Railways. The winters of 1960-62 saw the start of seasonal closures of some of the Manx lines, but in the summer of 1966 none of them reopened. The following year the system was leased to the Marquess of Ailsa, but the Peel and Ramsey lines closed for good in September 1968, the Port Erin route only staggering on with government help, which looked like drying up after 1974. In the following season only the very short stretch between Port Erin and Castletown operated, but in 1976 the service was extended to Ballasalla, while the former bus depot at Port Erin was opened as an excellent railway museum. Finally the remains of the system were taken over by the government in 1978 and the whole of the Douglas-Port Erin line is now operated as a seasonal tourist attraction.

Meanwhile the Groudle Glen Railway had sunk almost without trace. The zoo had closed in World War 2 when the last of the animals — the sea lions — were released, and, following the run-down of the other attractions in the glen, when *Polar Bear's* boiler tubes corroded through in 1962 the railway closed. That locomotive and a coach were shipped across to Sussex, and in due course were to be restored for operation at Amberley Chalk Pits Museum, but much of the track was lifted, and nature took over.

However, in 1982 restoration plans were announced by a preservation society, Groudle Glen Railway Limited, which started work on the formation and track as well as acquiring suitable rolling stock and motive power. A pair of Hunslet diesels was purchased from Dodington House, and in December 1983 one of these operated the first public service on the line for 21 years, some 700 passengers being taken to visit Father Christmas' grotto — the temporary engine shed! *Sea Lion* had been bought privately and, after various moves, was restored to full working order by the British Nuclear Fuels apprentices at Sellafield, and returned to the line in the autumn of 1987, 18 months after it had been formally reopened to the public.

Anyone frequenting the beach at Douglas during the season cannot fail to be aware of the clip-clop from the passing hooves of the Corporation's horse-trams. These operate along double track laid in the centre of the wide promenade between the Sea Terminal and the horse-tram sheds at the northern end, just short of the Summerlands leisure complex. Most of the vehicles are of the toast-rack variety, but there are various different designs, and much use is made of them for promotional purposes. A splendid new double-deck car (No 18), for instance, was constructed for the 1989 season, advertising Okell's Ales. No 44 is the Royal car, dating from 1907, and was actually driven by the Duke of Edinburgh during his 1970 visit to the island.

Travel on the trams is quite brisk, with the horses operating at a smart trot between stops, while the conductor takes the fares from the running-boards. The driver has a powerful handbrake available, operated by a well-polished brass crank-handle. The trams are operated 'on sight', but a special colour-light system is necessary to separate them from ordinary road traffic. This takes the form of a low, angled, two-aspect signal short of the main traffic lights at certain intersections, which gives the trams several seconds' start before the main lights turn to green.

Both the island's two electric railways are double-track throughout, with the supports for the overhead wires located between the tracks. The Snaefell line has simple insulators clamped to the cross-arms, while the arrangements on the coast line are much more elaborate. On the other hand, the Manx Electric Railway's cars can operate with a simple trolley pole, but the Snaefell Mountain Railway's cars have to be equipped with two bow collectors. The height of the latter's contact wire above the rails increases at each insulator, to prevent each bow rubbing on the fastenings as it passes. The continuity of the power supply is maintained because the bow at the other end of the vehicle remains in contact with the conductor.

The cars on the Electric Railway are of several different designs, and in recent years they have appeared in a number of splendid historical liveries. The first vehicles had exposed driving positions, but the later 'Winter Saloons' provided full protection for the drivers. However, the final 1904 and 1906 builds not only reverted to open cabs, but were of the toast-rack design, although roller side-shutters could be pulled

Right:
Despite their minimal decorative detail, the Beyer Peacock 2-4-0Ts in the Isle of Man have graceful lines. No 4 *Loch* poses for the cameras in the 1989 summer sun. *P. W. B. Semmens*

down if necessary. The majority of the trailers are also of open cross-bench design, reflecting the tourist-orientated nature of the line's operations.

Leaving Douglas, the Manx Electric Railway climbs the headland on the landward side of the road that was built as part of the original Howstrake development. It continues in the same fashion over the first summit and down to Groudle Glen, which is crossed on a fine arched viaduct, shared with what was originally a toll road. A mile or so further on the line crosses the main A2 road diagonally, and this is fully protected with a flashing-light system appropriate to a 'heavy-rail' open crossing. Coming south it is possible to watch the initiation of the warning sequence of the road lights as one's tram approaches, the yellow preliminary warnings giving way to the flashing reds before the corresponding white light that gives the tram driver authority to cross comes into view.

For much of the way from Douglas to Ramsey, the line runs close to the sea, and the views are much the best on that side of the line, which nevertheless frequently has to box the compass to maintain the 1 in 25 ruling gradient. At Laxey it runs inland up the valley for a mile or so to the junction with the Snaefell line, situated just north of the unusual curved viaduct across Glen Roy, built in stone with brick crenellations. Another viaduct beyond the interchange station, with its mixed-gauge track, takes the line across the Laxey Valley to return once more towards the sea. As the tramway turns northwards once more at Laxey Head, it reaches the most spectacular of all its clifftop locations. Climbing steadily around the succession of reverse curves, it finally reaches an altitude of 588ft above the waves breaking at the foot of the Bulgham Bay cliffs far below.

This is the summit of the line, which then keeps further inland for the remainder of the mostly-downhill stretch to Ramsey. Even these stretches provide some splendid scenery, especially in the spring, with carpets of bluebells in the woods, and the fields outlined with bright yellow hedges of gorse. As Ramsey is approached there are fleeting views of the seafront and pier, with the low cliffs beyond stretching all the way to Point of Ayre, the northern extremity of the island. A couple of hundred yards before the terminus the line passes the Ramsey Electric Railway Museum, which is situated in a large green shed, and is well worth a visit.

While the Manx Electric Railway runs to a published timetable, the Snaefell line operates solely on demand, the number of passengers being affected by the weather. This can change dramatically, not only from hour to hour, but over the four miles between Laxey and the summit. On one journey I made in May, it had been warm enough to wander around Laxey at midday without a coat. Just over half an hour later at the summit, visibility through the freezing mist was down to less than 50yd, while the deafening roar of the wind through the radio masts was unceasing!

There are six passenger cars on the line, and they operate singly. While they retain their vintage character, their electrical systems were completely renewed in the late 1970s, with the assistance of London Transport. Some second-hand traction motors were used from redundant Aachen trams, which enabled rheostatic braking to be provided. The Fell system is still retained for emergency braking, and also to keep the vehicles close to the track should there be a derailment. In some places on the lower stretches there are spectacular drops alongside the line, and, as an added precaution, right-hand running has always been used throughout, to keep descending cars further from the edge as they twist down the 1 in 12 gradients along the south side of the Laxey Valley.

Leaving Laxey for the summit, the best views are to be obtained on the right-hand side, the vista opening up as soon as the car has crossed the main road. This first section is single-track, but the line becomes double immediately beyond, after which the Fell rail starts. There is, however, a break in this by the junction to the car sheds on the left, which formerly had a complicated

Above:
Handsome Beyer Peacock 2-4-0T No 12 *Hutchinson* poses outside the depot at Douglas before taking up duty in summer 1962 on the Port Erin line.
Colin Boocock

Below:
The Isle of Man Railway purchased the two latest diesel railcars from the County Donegal Railways when the latter closed in 1960. The pair, which operated singly or with trailers when in Ireland, were used in back-to-back formation on the Isle of Man. The leading car hauled its neighbour dead, in whichever direction they worked. No 19 leads No 20 in this flashback to 1962 on a Peel to Douglas working. *Colin Boocock*

section of Fell rail which slewed across with the track. Nearby is another small shed, which houses the second form of motive power used on the line. Since 1951 the Air Ministry, and now the Civil Aviation Authority, have operated an important radio station at the summit, and to get their staff up and down the line they have their own Wickham railcars, also fitted with Fell brakes, which are used when the public railway is not running.

As the line begins its climb, a splendid view is obtained of the red and white *Lady Isabella* water wheel, situated across the valley. Completed in 1854, this has a diameter of 72ft 6in, and until 1929 was used to pump water out of the nearby mines. Now the property of the Manx government, it has been restored to operating condition, and from the Snaefell Railway can be seen turning, while the slow oscillations of the huge flat-rod transmit the thrust of the crank to where the pumps used to be situated.

When the Snaefell line was built, it had its own power station, situated right beside the line, two-thirds of the way to the summit. The necessary coal had to be taken up by train, and the flue gases passed under the tracks before being discharged up a tall metal stack. Later it was found easier to transmit power by an overhead high-voltage cable, and the generating equipment was replaced by a simple substation, but the chimney's foundations can still be seen.

Not far above the old generating station the line reaches the level crossing with the A18 main road at the Bungalow, situated on one of the fastest stretches of the Tourist Trophy motorcycle course. With high-performance bikes lapping this at well over 100mph, it is clearly not possible to continue to work the railway over the crossing while the races are on, but an interesting alternative has been adopted. One car stays on the upper section, and shuttles between the Bungalow and Summit stations, while the others work the longer stretch to and from Laxey. A temporary footbridge over the road enables passengers to transfer between the two halves of the line.

Bungalow station is situated on a col, and the line then crosses over to complete its climb to the summit by means of a clockwise spiral making virtually one complete circuit of the mountain. On a clear day the view from the summit is superb, with Wales, Scotland, England and Ireland all visible, but even when there is no incentive to admire the scenery, the hotel provides shelter and food, as well as selling souvenirs. The cars reverse outside it on a short length of single track, a very unusual single-bladed point being provided at the Laxey end. The point lever has to move the toe of the blade almost the full 3ft 6in between the rails, and the horizontal lever looks more like a capstan bar than something associated with railway practice.

The terminus of the steam railway is situated inland of the harbour area. It is bounded on the north side by a steep, tree-lined cutting, while behind the bufferstops are the twin gold-capped minarets that crown the passenger entrance, from which a long flight of steps leads down to the well restored, brick-built station offices. Inside there is an attractive buffet, and bookshop, where one can purchase the modestly-priced A5 booklets about the different railway systems on the island. They provide a lot of useful information even if the one describing the Snaefell line carries on rather excessively about the author's objections to the 1970s' conversions.

The station has three platforms, the more southerly pair being an 'island', at the end of which used to stand the well known pair of bracket starting signals. These were replaced many years ago with colour-lights, but they, too, have been superseded, this time by radio signalling. However, at the time of my 1989 visit it was hoped that it would be possible to reinstate the semaphores to help recreate the vintage atmosphere. The signalbox, with its unusual locking frame, is now completely separated from the running tracks by part of the bus parking area, and the old corrugated-iron carriage sheds behind it are used to store locomotives, rolling stock and buses.

The workshops and locomotive running shed are situated beyond the platform-ends, on the other side of the line. One of the programmes in the 1988 BBC TV series *The Train Now Departing* showed the craftsmanship that goes on here to maintain and rebuild steel locomotives and wooden coaches, the staff again doubling up to help work the trains during the summer traffic peaks. As already mentioned, all the locomotives built for the Isle of Man Railway proper were 2-4-0Ts. Although there are many detailed differences between individual locomotives, their common Beyer Peacock ancestry is very obvious, all the

Left:
The Douglas horse trams are a major attraction to transport enthusiasts. The conductor is seen on the footboard collecting fares as the tram makes its way along the sea front towards the ferry terminal. The fares quoted on the tram front (6d all the way, 4d stage) were current in 1958 when this photograph was taken. *Colin Boocock*

way from their copper-capped chimneys to the inclined cylinders set above the outside springs and axleboxes supporting their pony trucks.

The IoMR had a total of 15 of these locomotives, and, one way or another, an extremely high proportion of them still exist. Four are needed to maintain the seasonal services (three working and one spare), with another undergoing long-term overhaul. Two are on exhibition in the museum at Port Erin, together with the 0-6-0T *Caledonia*. Although two of the four Manx Northern Railway locomotives disappeared long ago, only one of those from the older system (No 2 *Derby*) was scrapped. All the others are still extant, although No 3 *Pender* now only exists in sectioned form at the Greater Manchester Museum of Science & Industry. Four of them are owned by the Isle of Man Railway Society, and the restored frames of No 7, *Tynwald* are on exhibition at Castletown.

During my visit in May 1989, *Loch* (No 4), *Hutchinson* (No 12) and *Kissack* (No 13) were at work on the Saturday morning, while *Maitland* (No 11) was still in the works having the plating of its side tanks renewed. The three active locomotives were in different red, blue and green liveries, and had been turned out in splendid condition, with all their brass and copperwork gleaming in the spring sunshine. I was in the island with a party from the Railway Study Association, and we had chartered a special train for the round trip to Port Erin, for which *Kissack* was provided, together with the red and white three-coach bar set which is regularly used for such workings. The serving area is in the centre vehicle, which also includes a splendid blue leather seating area, while the outer coaches were of open design. Each seating bay is provided with a small table for passengers' glasses, the overall result providing a very pleasant way of seeing the southern part of the island.

Leaving the outskirts of Douglas, the railway climbs for the first five miles to the 264ft summit near Ballacostain. The sea soon comes into view away to the left, together with the end of the marine drive. Until 1939 this was served by the Douglas Southern Electric trams, whose other terminus was reached from the harbour side by a funicular. The countryside is far less rugged in this area, but, even so, ranges of hills are always visible in the distance. The line snakes its way past farms and villages, traversing level crossings manned by local part-time keepers. After seeing each train safely across they take off their high-visibility jackets and return to their other business or leisure activities, providing another example of the way in which the cost of running the line is kept to a minimum.

Many of the later intermediate station buildings have been sold, but the southern terminus at Port Erin has, like the one at Douglas, been extensively restored, with many interesting old signs and posters in the waiting room. On the other side of the tracks is the railway museum, already referred to, with its extensive collection of full-size exhibits, as well as displays of posters, tickets and photographs. One item that provided an interesting commentary on the status of the railway in Edwardian times was the poster announcing the extra train services to be run in connection with the Great Naval Sports, held in Kirby, near Douglas, during the visit of the Atlantic Fleet in August 1908. Third-class returns from the terminus only cost 4d ($1\frac{2}{3}$p).

The importance of the railways to the Isle of Man's present-day economy is demonstrated by the appointment of a government minister with responsibility for tourism and transport, with separate chief officers responsible for the two different halves of that portfolio. The remit for the second of these is to provide an *integrated* transport system for the benefit of residents and tourists, which forms a very marked contrast to the situation in mainland Britain.

To add to the appeal of the vintage atmosphere of the two railway systems are some extremely attractive rover ticket offers. For instance, in 1989 adults could purchase a seven-day railway rover for only £12. The first child in the family travelled for half price, but up to three more could then go free. Because of the limited capacity of the Snaefell line, a maximum of only one round trip was permitted over this on any of these particular tickets. Market research has shown that comparatively few local inhabitants use the railways or are aware of their tourist potential, so a special rover ticket — £10 for the whole season in 1989 — is available for them.

So, for the residents as well as the holidaymakers, there is every justification to travel on these splendidly preserved and restored railway systems, which still fulfil their century-old role in the tourist industry.

Left:
The 2-4-0T *Sea Lion* takes water at the end of its run on the Groudle Glen Railway. This was the scene in 1958, a few years before the railway closed. Happily, *Sea Lion* is back there now and the railway sees new life in its restored status. *Colin Boocock*

Photo Feature: Modern

Peter J. Robinson

The modern railway in Great Britain is becoming ever more colourful!

Each British Rail business Sector now has its own livery, or sets of liveries. BR is getting better at reducing the incidence of trains of mixed liveries, though some remain while stock has yet to be repainted in the latest colours. Nowadays the humble freight train splashes colour around some of the less colourful parts of the country.

These four pictures show how the use of striking colours enables the modern train to stand out from its surroundings.

Left:
On 22 March 1987 IC125 power cars Nos 43040 and 43043 take the 11.35 Edinburgh-King's Cross through the Berwickshire countryside.
Peter J. Robinson

Traction Action

Below left:
The cross-country sub-Sector of InterCity received much cascaded Mk 2 stock and for a time found it difficult to keep its train sets in consistently liveried rakes. Electric Bo-Bo No 86250 *The Glasgow Herald* heads a typically mixed formation on the 08.34 from Poole to Glasgow and Edinburgh on 25 March 1987. *Peter J. Robinson*

Right:
Network SouthEast's train liveries broke with the railways' former aesthetic tradition. On 8 August 1989, Class 50 Co-Co diesel electric No 50017 *Royal Oak* takes the 09.15 from Waterloo to Exeter past Coker Wood near Yeovil Junction.
Peter J. Robinson

Below:
Railfreight Coal sub-Sector Class 37 Co-Co No 37167 brings the 02.30 Speedlink coal train from Mossend to Healey Mills past Stonebridge, Durham on 26 May 1989. *Peter J. Robinson*

Semmering — The First Railway Across the Alps

Colin Boocock

It is very peaceful. The only sounds are the distant squawking of birds, the rustle of a light breeze in the pine trees above us, and the trickle of a fast-flowing stream way down at the bottom of the valley. The bright morning sun casts deep shadows among the rock faces of the mountains around us. Occasionally, we hear a metallic 'ping!' as a rail on the track beside us expands slightly in the summer heat.

Then a distant rumbling is heard, lower in the valley. It swells and fades. As the noise grows louder we can distinguish the sharp exhaust of a steam locomotive, sometimes slow, sometimes faster — then we clearly hear two exhausts together but at different speeds. Two columns of smoke appear round the edge of a hill in front of us — and then disappear!

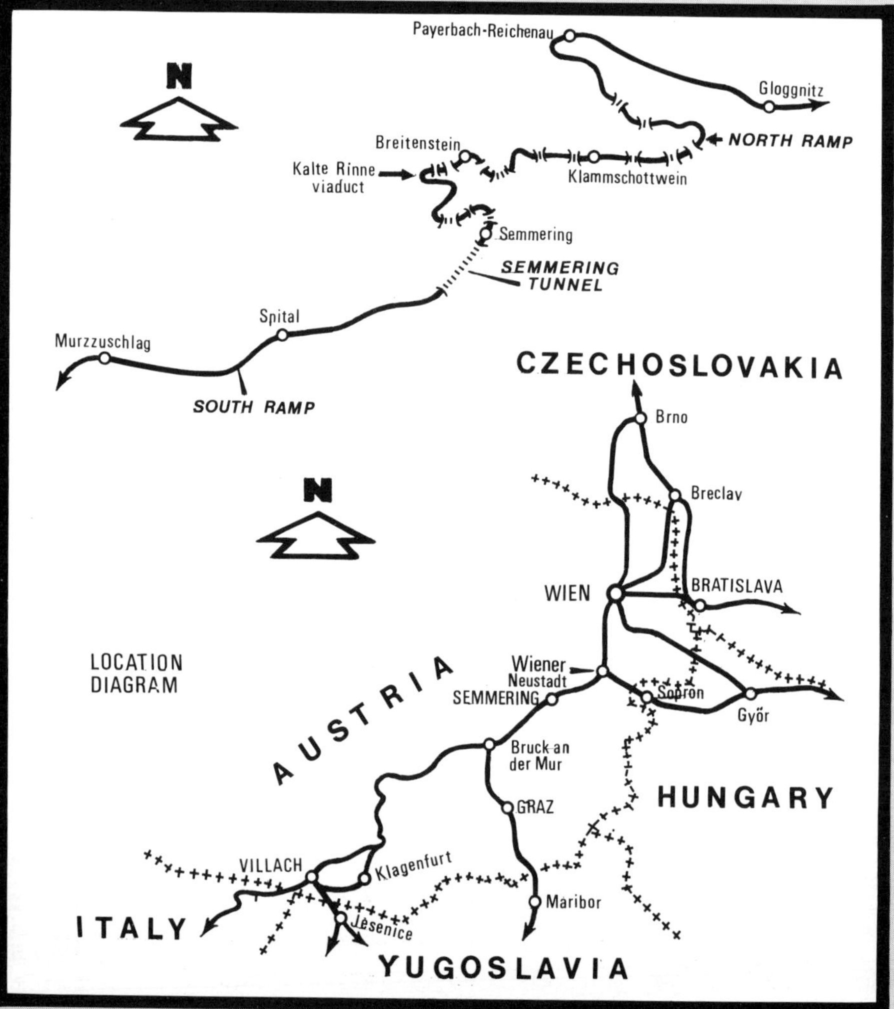

With a double burst of sound and smoke the train emerges from the tunnel opposite us across the valley, the two leading engines now panting across an ornate, double-deck, stone viaduct as they curve towards us, followed by 9, 10, 11, 12, yes 13 dark green or brown main line carriages. Another engine follows, working hard at the rear. Although the gradient approaches 1 in 40, the engines' wheels are revolving moderately rapidly. The front locomotive is nearing us now, a tall, gaunt, black, slab-fronted 2-10-2T with its narrow Giesl ejector chimney sending its exhaust sky high. The big 4-8-0 behind it is clearly playing its part in maintaining the train's 30mph progress. As the crackle of the two exhausts fades around another corner we try to read the destination boards on the sides of the carriages as they pass by: Graz, Villach, Marseille, Trieste, Roma. Before we can take it all in, the banking engine approaches. It is another large tank engine, also of Class 95, its 10 coupled wheels pounding away purposefully as it helps this international train towards the summit at Semmering. The loud chattering of the banking engine's exhaust fades as the train disappears from view. The sound of steam gradually merges with the rustling of the trees, and soon peace descends again.

The experience just described belongs to little over 30 years ago. The Semmering route linking Wien (Vienna) with the Adriatic coast is historic for a number of reasons. It was the first of the great Alpine railway crossings to be opened (in 1854) and the last to be electrified, in 1959. From the outset the railway was adhesion worked with no recourse to rack or rope working. The Semmering line is also outstanding for its scenery. Its 105 years of steam operation saw locomotives with six, eight and 10 coupled wheels. The last 12-coupled locomotives to work on main lines in western Europe worked on the Semmering as late as 1956!

The route owes its origin to the desire to link Wien with the Adriatic port of Trieste, both centres then being in the territory of the Austro-Hungarian empire. Any route

Left:
Map of the Semmering railway crossing of the Alps.

Left:
Two Class 42 'Krieglok' 2-10-0s haul a heavy freight up the north ramp of the Semmering line in the mid-1950s. As they cross the Krausel Klause viaduct it can be seen that the load includes a new Floridsdorf-built broad gauge locomotive for the Indian Railways, carried on low level transporters. This illustrates the size of the Austrian loading gauge. *H. Navé*

Below left:
Class 95 2-10-2T No 95.114 approaches the rear of an express train it is to bank over the Semmering line, at Gloggnitz on 9 September 1958. *Colin Boocock*

Bottom left:
Tall, but quite handsome, were the 4-8-0s which hauled the express passenger trains over the Semmering line for several decades up to the late 1950s. No 33.114 pauses at Bruck an der Mur before tackling the south ramp on 9 September 1958 with an express from the Balkans to Wien. *Colin Boocock*

between the Danube valley and the Adriatic faced the need to cross the Alps. Albeit not the highest in the Alpine range, nonetheless mountain passes such as the Semmering are well over 1,000 metres above sea level. To drive a main line railway through such country in the 1840s and 1850s was truly a daunting task. Engineer Carl Ritter von Ghega, a man of outstanding ability and vision, was appointed to plan and lead the project.

The bulk of the work was undertaken in the remarkably short interval between 1848 and 1853, so useful was the project as a means of alleviating the unemployment of the time, and so important a link did it promise to be. The route over the Alps was opened in May 1854, but the objective of linking Wien with Trieste was not to be achieved until 1857 when all sections of the route had been opened. Celebrations were therefore delayed until 1857 when the complete project could be commemorated.

Over 130 years later, the journey from Wien to Graz over the Semmering route is still one of Europe's most striking and attractive. The modern-day traveller could leave Wien Südbahnhof in one of ÖBB's excellent former 'TransAlpin' five-car push-pull sets; or he could leave in a locomotive-hauled train, in which case a Class 1042 or 1044 electric locomotive could be hauling clean modern coaching stock, all liveried in shades of peaches and

Right:
The striking, double-deck stone viaduct over the Kalte Rinne stream is a superb example of the architecture of the railway across the Semmering Pass. On 13 September 1987 a pair of Class 1042 Co-Co electric locomotives, Nos 1042.581 and 674, cross the viaduct upgrade with a heavy freight. *Colin Boocock*

cream. The electric system is the standard German 15kV, 16⅔ cycles ac with overhead collection.

For the first hour speed is unrestrained as the train heads south and southwest, stopping midway at Wiener Neustadt. As the mountains begin to envelop the railway, the train reaches a small industrial town with the unforgettable name of Gloggnitz. This used to be the station at which the steam locomotive hauling the train would pause to take water while pilot and/or banking engines were attached. Thirty years ago the large locomotive depot here closed, as the need for steam locomotive assistance of heavy trains vanished under the influence of the electric wire.

Today, our single electric locomotive halts only briefly with its train. We glimpse a preserved Class 95 2-10-2T on a plinth by the station as our train leaves and heads towards the mountains, along a valley with seemingly no way out ahead. The mystery is solved as soon as we have swept through Payerbach-Reichenau station, for the train curves swiftly left, over a grandly designed stone viaduct. Almost doubling back on our route, the train begins to climb up the opposite side of the valley. Speed settles down to a surprisingly rapid pace as the double track railway gains height, curving sinuously. Soon, low on the valley floor to our left, we see a town against the sun and we realise it is Gloggnitz where we stopped less than 15min or so previously.

The train curves south now, hugging the mountain face, diving in and out of brief tunnels and over viaducts whose short length belies their height in space. Most of the tunnels have ornate mouths but are rough blasted rock inside (some have more recently been lined with concrete. At intervals loop lines peel off the tracks as the train rushes through stations, each open to the elements except for the large station buildings. Often there is a small, two-storey railway house by the lineside. Many of these are favoured as standpoints for railway photographers.

The train is now running on a ledge cut into an almost sheer mountain wall. The railway curves as a contour line must, breaking away into tunnels and on to viaducts where no other course is possible. After passing Breitenstein, speed slackens and the train emerges from a short tunnel curving sharply left on the most famous viaduct on this line, Kalte Rinne. Looking back we see how this viaduct's double deck construction of Romanesque arches, created by Ritter in stone on a horizontal curved plane, give the Kalte Rinne viaduct its almost surrealist and most striking appearance. Behind in the near distance rise the Rax mountains. Pine trees almost envelop the viaduct's lower parts. Such magnificence can be admired at length by anyone willing to alight at Breitenstein, follow a wooded path to the valley floor, and then walk up a road which passes underneath the viaduct.

This is not so from the train, however. Only fleeting glimpses of Kalte Rinne are possible as the railway winds between banks of trees, and then curves sharply away to the right. Soon, on a hillside to the front, stands another monument to this lovely railway, the Südbahn Hotel. This 19th century colossus was set up to enable the wealthy Viennese to rest in the clear and healthy mountain air, in good walking and climbing country.

The train swings to the left now, and through another tunnel, and soon curves to where the tracks fan out into Semmering station. Here we can alight and look around. Ahead are the two portals of the twin tunnels, the first opened in 1854, the second in 1952.

In the period after World War 2, traffic on this route began to expand rapidly, yet the railway's fabric was worn and decayed. An expensive programme of renovation of tunnels and viaducts was undertaken. The condition of the original Semmering Tunnel was threatening instability, so a second, single bore was driven. Diversion of traffic into the new tunnel enabled the original bore to be temporarily closed so that engineers could spray a thick concrete lining inside; the older tunnel was then reopened as a second, single-line tunnel. Track and signalling renewals were followed by electrification, and a new era dawned.

At Semmering station there stands a tall stone monument to Engineer Carl Ritter von Ghega, a tribute to the high standing which this great man and his railway brainchild attained. His farsighted vision and excellent engineering provided a railway alignment capable of adaptation to modern railway traffic.

As is so often the case with railways which climb over mountains, the run down the other side is different in character from the climb. The 'south ramp' of the Semmering is not so steep nor as curvaceous as the 'north ramp'. It nonetheless posed a challenge to heavy freight trains in the days of steam, and the ÖBB drafted here its most powerful freight locomotives. Allocated to the depot at Mürzzuschlag at the foot of the south ramp, these big 2-10-0s were often accompanied by the attractive former Württemburg 2-12-0s as banking engines. Nowadays, freights are headed by pairs of Bo-Bo electric locomotives which scale these inclines at line speeds undreamed of even just after the last war.

That is just as well. The expansion of traffic over the Semmering has continued well after electrification. The intrepid observer who nowadays might stand on the hillside overlooking Kalte Rinne is kept entertained virtually all day by a succession of heavy freights, interspersed by the surprisingly frequent passenger trains which pass by.

Much of the freight over Semmering is using Austria as a land bridge from the Comecon countries of Poland, East Germany and Czechoslovakia to Italy and Yugoslavia and vice versa. Southbound trains of brown Polish coal pass northbound loads of steel pipes. Trains of tankers follow mixed freights of considerable length. Yet two Bo-Bos are all that is required to haul the heaviest freight trains up the north ramp, so competent are these 5,000hp and 7,100hp locomotives.

Passenger services are grouped around a basic hourly fast service from Wien to Bruck an der Mur, which then heads variously to Graz or Villach. These trains are normally rostered for five or 10-car push-pull sets, the pride of this line. It is difficult to remember that these beautifully styled and well-appointed multiple-unit trains are now 20 years old. They are always immaculate, even their light cream roofs being kept clean. The ÖBB is justly proud of them. Other trains include many heavy international services between Wien and Beograd (Belgrade), Sophia, Bucharest, Milan, Trieste and Rome, and even the south of France. Lastly there are the stopping, or Regional, trains, always a single Bo-Bo with four to eight peaches-and-cream-liveried carriages, as pretty a sight as one could expect on a modern, electrified railway.

Capacity is still a problem. The continued development of international trade leads to more freight and more passengers. Somehow the downhill transits of the pilot electric locomotives released at the summit have to be fitted in between revenue-earning trains. Despite two-way signalling which allows trains to run right-hand instead of left-hand when necessary to pass or overtake, more capacity is going to be needed.

Thus we read of the proposal to build a base tunnel, bypassing Semmering altogether. Four of the five proposed alignments start near Gloggnitz and tunnel virtually straight southwestwards, avoiding the curves and tunnels of the north ramp. The fifth alignment takes a more northerly route through Payerbach. Most bypass Mürzzuschlag before rejoining the existing route.

Above:
An Austrian Federal Railways push-pull electric multiple-unit express train of Class 4010 crosses the Kalte Rinne viaduct with the 12.00 from Wien Sud to Graz on 13 September 1987.
Colin Boocock

Austria is not an exceptionally wealthy country, though it is climbing the economic scale. The timing of the base tunnel decision will almost certainly be dictated by a mixture of economics and politics, bolstered by the imminent sheer necessity of increasing line capacity.

We may not therefore have a great many more years left in which to enjoy the journey on Carl Ritter's railway up the north ramp to Semmering. The experience, on a sunny day, of leaning against the semi-opened window of a Regional train making its way through this rugged Alpine fairyland is too good to be missed. Go out of *your* way. Find an excuse to ride the Semmering line from Gloggnitz to Mürzzuschlag and traverse its 16 viaducts and 15 tunnels before it is too late!

Today's Men of Main Line Steam

Joan Jackson

Picture the scene at the platform end of a busy main line station. A hundred or more passengers and well-wishers watch expectantly as a plume of smoke heralds the arrival of a steam locomotive, which is to work the special train already in the platform. Station staff try unsuccessfully to hide their interest under a pseudo-blasé exterior.

But what is this attached to the locomotive? Another coach to be added to the train? Surely it is long enough already. As it draws nearer, it is obvious that this is no ordinary coach, being of a different colour from the rest of the train and with people already inside. Many passengers comment on the coach, which seems to be out of bounds to them. So who is it that should have the privilege of riding in the prime spot behind the locomotive?

Eight men and women are nominated to ride in the locomotive support coach: a responsible officer, engineer, two sales officers and four representatives of the society, which may include the locomotive owner, where the locomotive belongs to an individual. These eight people have all passed a BR rules exam and are familiar with the working of their locomotive.

So meet today's men of main line steam, who could well be your next-door neighbour, bank manager, doctor, lawyer, writer, mother, engineer, sailor or even railwaymen on a busman's holiday. As these volunteers come from all over the country to work on their chosen locomotive, a home from home is required that will act as a base for living and sleeping, eating, locomotive maintenance and administration.

Consequently the ubiquitous support coach is now to be seen on all steam-hauled railtours. The coach enables the locomotive society/owner to operate all over the country in comparative comfort, as the group lives hundreds of miles from home for days on end.

In the early days of steam operation, conditions for support groups were much less organised. Picture again our scene at the main line station but several years ago. Mingling with the passengers craning their necks for a glimpse of the locomotive, would be a number of people in overalls with assorted bags and sacks containing rations for the day, tools and oddments, having been transported there from the locomotive stabling point. Satisfied that all is well on the footplate, the volunteers scramble themselves and kit into the leading coach in search of a vacant brake van or vestibule end where they could squat for the journey, emerging at stops to throw coal forward for the fireman, water the locomotive, search for an open buffet and change places on the footplate. Such conditions were not desirable and were not to last.

With main line steam enjoying a strong revival during the last decade, over an expanding area of approved routes, the locomotive support groups have had to face daunting logistical problems in order to maintain a viable service to the fare-paying public. The arrival of the coach on the scene was a godsend for this purpose. Not only could the group sleep in comfort in the compartments after a wash and change but cooking facilities were installed and a workshop with bench and fittings meant that running repairs could be undertaken.

All of this makes for a much more professional approach to locomotive care and maintenance. Such an approach is vital as the scope of the steam scene has changed from when many enthusiasts actually preferred a dusty locomotive as being somehow more authentic, to the era of the upmarket dining special, when the passengers quite rightly appreciate an immaculately turned-out engine as part of a truly first-class experience.

In order to present this glamorous image for the few hours in the day that the locomotive is on view to the public, there are hundreds of man-

Below:
There are no coaling towers or high level coal stages here! The Merchant Navy Locomotive Preservation Society's No 35028 *Clan Line* is coaled the hard way at Stratford-on-Avon on 12 April 1987. *Keith Jackson*

Right:
Class 5 4-6-0 No 5407 blazes a black trail over Ais Gill Summit with the southbound 'Cumbrian Mountain Pullman' of 24 August 1983.
Keith Jackson

hours of unseen preparation, often carried out in cold, wet, cramped conditions. For example, when 'Black 5' No 5407 first went to Scotland to work the West Highland trains, there was only one regular to look after the engine; his chief partner had a job to go to and volunteers from Steamtown went up to help during their holidays.

Based at Fort William with a couple of sleeping cars on loan from InterCity, conditions were primitive. The engine had travelled to Mallaig with the rods off and it had to be ready for action within hours. Fish and chips was the order of the day at Mallaig Junction and the West End Hotel was the venue for evening meals.

On shed there were few facilities for steam engines. The fire had to be paddled out into a drum on a trolley. Things have improved since the service became an annual event but it is hard work to run an engine several days a week for weeks on end and, inevitably, there is maintenance to be done. Also it is expensive living away from home and eating out in restaurants. As time has gone on, No 5407 has acquired a coach and a larger workforce, including a cook, so they are ready for anything! As the running proliferated and midweek trips became the order of the day on the Settle & Carlisle, West Highland and Crewe-Holyhead routes, support crews found themselves operating at full stretch to provide sufficient well-qualified members to service their locomotives. Holidays have to be sacrificed, often at the expense of the family, or shifts changed at short notice and much overtime worked to compensate colleagues for filling in.

In the case of the North Eastern Locomotive Preservation Group, which at one time had 'Black 5' No 5305 in Scotland and *Sir Lamiel* in the south, the problems of manning and maintenance are alarming. Thousands of miles of commuting are necessary to keep both engines on the road. The more popular the locomotive, the more the demands made on the team. For obvious reasons, BR requests the reliable engines with a proven record and a competent support group for its own operations; particularly the more prestigious ones. This demands a full commitment over the whole year.

With the high cost of bringing a locomotive to BR main line standards, it is essential that the engine is kept ready for action as much as possible and this means that the support crew must be on call. In some cases, this has meant that steam is the priority for the majority of the time, one's life being work and the engine. There is no room for those who can manage to find time to come and play trains when the engine is on the main line but are too busy to come and help prepare the locomotive or spend weary hours grappling with heavy, awkwardly-placed greasy parts during overhaul.

Many volunteers come along to help in the knowledge that they are unlikely to get a ride behind their chosen engine unless they can afford to pay the train fare. When No 4498 *Sir Nigel Gresley* first came to London in the bitterly cold January of 1985, at first a couple of regulars who could spare the time, disposed of the locomotive after the run and cleaned it. There was considerable excitement at Marylebone and various offers of help were forthcoming, one fireman helping to clean the smokebox and going on to have several firing duties with the engine.

Many London-based members of the A4 Locomotive Society, delighted to have the engine in the south, banded together to form a cleaning gang and clean it for each trip; this relieving the main workforce from the Northeast of a big chore. It enabled them to concentrate on mechanical maintenance and saved hours of their time.

Obviously with ageing machinery, faults occur. Most of the Pacifics working out of Marylebone have suffered a broken spring at some time. When this happened to No 4498, a replacement was sent by train from Carnforth and had been fitted by midnight the same day! *Duchess of Hamilton* borrowed a spring from *Duchess of Sutherland* and *Clan Line* borrowed from *Blue Star*.

The team has to be prepared to tackle all sorts of jobs, with the emphasis being on preventive maintenance. A comprehensive selection of expensive tools has to be carried in order to effect repairs as necessary. Safety is of paramount importance to the continuation of main line steam, so care of the locomotives is vital. A great deal of pride goes into turning out a fine locomotive for BR crews to handle.

The Gresley and Bulleid Pacific support crews are always anxious to see if the fireman will maintain the economical classic thin saucer-shaped fire with the back corners made up, that so suits their locomotives. Thankfully, as the locomotives travel the country, an increasing number of BR crews are becoming accustomed to strange locomotives and go to great lengths to familiarise themselves with them.

Obviously, it is a great privilege and wonderful responsibility to represent the group on the footplate on BR main lines and it certainly compensates for all the hard work. It is a real joy to observe an expert BR crew handle the locomotive to maximum effect with maximum economy. In a small minority of cases, where a crew is inexperienced in the type of locomotive a fireman will pile coal into the firebox on a steep bank and it is disappointing when the inevitable slip or stall ensues. Not only is the coal bill ridiculously high but the unnecessary cost of obtaining a diesel to get the train out of trouble, the inconvenience to passengers and the unfair slur on the locomotive and its support crew does the steam scene no good at all. Fortunately, this is rare and the 3hr-plus marathon of cleaning the resultant clinkered mess at the end of the day, is not often experienced. Most groups have tackled such a mess at some time. This is bad enough over a pit at the home base but it is no joke at all at a strange depot in the dark and pouring rain to have to paddle out a huge fire, with no hope of a pint at the end of the day to relieve the gloom!

A typical trip starts a few days before the run when the working party arrives in time to prepare the locomotive and coach for BR inspection. If any jobs need doing before the engine is in steam, they are tackled straight away or, if a job needs to be checked before steam is raised, then a slow fire is started after making sure that there is water in the boiler, all washout plugs are tight, smokebox and grate are clean and the ashpan empty. About half a glass of water showing in the gauge glass is right for lighting up, which gives room for expansion as the water heats. The day before the run is particularly busy. Coaling often takes place now, which may mean filling a mechanical bucket and trimming the coal on top of the tender or, where the tender is of small capacity, as with *Clan Line,* bags must be filled by hand and manhandled on to the tender, with extra bags loaded on to the coach. It is all hands on deck for this dirty, heavy job. Other jobs are allocated. Someone will fill the tender and coach water tanks, whilst someone else fills the sandboxes. The locomotive, tender and coach brake adjustment is checked, engine and coach lamps are filled, oil bottles topped up and the locomotive and coach cleaned and the footplate prepared for the run. Where the locomotive is working from its home base with society members living nearby, most of the cleaning will have been done before the working party assembles.

The locomotive and coach will be presented to the BR mechanical inspector at a mutually agreed time within 24hr of the train departure time, for its mandatory examination. It is hoped at this time that no further defects would be noted. If any faults are found, the BR inspector and locomotive group engineer discuss which work must be done immediately and which can be safely left until after the run. With most of the preparation now done, the fire is banked up for the night, the group washes and brushes itself up and retires to the local hostelry for a hot meal and a pint, if all goes to plan and there is still time!

On the morning of the run, the person in charge of the fire gets up very early to start bringing it round ready for the trip. The rest of the group soon follows to carry out the final duties. This will include one very important person making tea and preparing breakfast. Oiling up will be completed at this stage, the lamps lit and put in position, tender water level topped up to maximum and any last-minute job completed. As final departure time approaches, with breakfast eaten, the fire is prepared in readiness for the BR crew to take over. They usually arrive about 20min before the departure time off shed to install themselves and billy cans of tea on the footplate.

At this stage, with the exception of one nominated person, who will ride on the footplate, today's men of steam retire to the support coach, hoping that all their efforts will be rewarded by a professional performance by the BR crew now in charge of their locomotive.

We are now back at that main line station, as locomotive and support coach back down on to the waiting train. Support group members pick out familiar faces amongst the crowd and exchange greetings before the intending travellers hustle aboard.

The group listens anxiously as the brakes are released and the train set into motion. This is a crucial moment for them as their work is now coming to fruition. The ever-increasing rate of the exhaust beat is like music to their ears as the performance unfolds like a symphony. With the journey now underway, the team can begin to relax and look out of the window, spotting friends and photographers at strategic lineside locations. Many of these will be seen several times before the day is out.

Meanwhile, the nominated sales team will have discarded their overalls in favour of smarter apparel and set out to walk through the train

Left:
Can you imagine the amount of hard work necessary to produce this high standard of presentation of a preserved, working locomotive? The support crew of 'Jubilee' 4-6-0 No 45596 *Bahamas* must have been proud of their engine's appearance at Coalville open day on 11 June 1989. *Hugh Ballantyne*

talking to passengers and selling souvenirs of their engine to raise funds to help keep it running. Many new members join the societies through these encounters, and with their subscriptions they play a vital part in funding the locomotive. This public relations exercise is demanding, time-consuming but very rewarding, even though much of the day's journey is lost to the people doing it!

As the journey progresses, water stops have to be made, which will result in the support group swinging into action as in the past. This time, however, there is no mad dash for the buffet. The kettle in the support coach is on ready for when they climb back aboard.

Most trips run smoothly, and all too soon it is journey's end and either the engine and coach depart as the train returns passengers to their starting point, or the end is at a terminus and passengers crowd round to have a last look at their locomotive before making their way home. With a bit of luck, the fire will have been nicely run down by the time the engine is safely on shed. As soon as it comes to a stop, the BR crew members depart having thoroughly enjoyed themselves, we hope. At this point, the support group takes over to dispose of and put its engine to bed. If it is not running the next day, the fire is dropped, footplate tools gathered together, lamps put away, boiler filled and controls securely closed. A quick look round the engine may be called for at this time to check that

Above:
Steam in the West Highlands looks idyllic. But the people who look after Class 5 4-6-0 No 44871, seen here working ScotRail's West Highland set of carriages past Loch Eil on a Fort William-Mallaig train on 25 May 1989, have to face the logistics of working a long way from home for the season.
Dr W. A. Sharman

Left:
On 'The Yorkshireman' railtour of 1 May 1989 Class V2 2-6-2 No 4771 *Green Arrow* heads south from Armathwaite en route from Carlisle to Appleby. The support coach is next to the engine.
Dr W. A. Sharman

all major components are secure and bearings cool.

By this time, some members may well have set off for home to take up their own jobs the next morning, leaving just a few to clean and tidy up the next day. When the engine is cool enough, the smokebox, grate and ashpan will be thoroughly cleaned out ready for the next run.

It should be appreciated that these procedures also have to be gone through for moving an engine and coach across country without a train. Regulars on the scene are used to these movements, which usually take place at extremely unsocial hours; often negotiating unusual routes after dark, in order to avoid disturbing the smooth running of the railway network.

It is at such times as these, that people involved are dependent upon having understanding employers, as BR cannot afford to take into consideration society members when it plans these moves. We are, after all, privileged to be permitted to run on BR main lines in the first place.

Thrown into the deep end of all this was Bert Hitchen and his support team. Having operated his locomotive, No 34027 *Taw Valley*, successfully since restoration on the Severn Valley Railway, the team was suddenly called upon to prepare its locomotive for BR main line running. This was to be a baptism of fire. They left Bridgnorth on a Wednesday for weighing at Derby on Thursday. On Friday they had a loaded test run from Derby to Sheffield and back with the intention of proceeding to Banbury to act as standby engine to *City of Truro*, hauling an InterCity VIP special from Didcot to Stratford-on-Avon on Saturday. This was to be their first main line commitment.

After a successful test run, the locomotive was en route from Derby to Birmingham when trouble developed with the middle big end. They proceeded to Tyseley, where inspection revealed that the big end brasses would need attention. This unfortunately meant that they were unable to take up their standby duties at Banbury.

Working throughout the night and day, they were able to take the brasses to Bridgnorth for repair and to get them back to be refitted to the engine in time for them to take up their booked path forward from Banbury back to Derby, to enable them to go on to Crewe to take their allocated work on BR's 'North Wales Coast Express' trains.

The crew found working on the main line quite different from a preserved railway, where distances between stops are less, trains are lighter, speeds are lower and there is plenty of help on hand. However, they soon got into their stride, tackling problems as they arose with skill and speed and went on to have a good first season in 1989; gathering experience and starting to build a reputation for their locomotive.

Even the most experienced support crews have their fraught moments. One such occasion was in April 1986, when the EMI recording company hired a train from London to Gleneagles, which was to be hauled from Edinburgh by two of the National Railway Museum's locomotives: No 4771 *Green Arrow* and No 92220 *Evening Star*. Manned by volunteers from the Friends of the NRM and Museum staff, the event was to take place on Monday 7 April. For several days beforehand, the volunteers prepared and cleaned the locomotives ready for their overnight journey from York to Edinburgh. This passed uneventfully with water stops at Newcastle and Berwick before proceeding to Millerhill, arriving at daybreak. From here on, things began to go wrong. Their arrival was unexpected, there were no crews rostered and they couldn't find any water! Members of the

Below:
No 5305 always glistens, such is its reputation and such is fact. On 28 August 1989 the mascot of the Humberside Preservation Group climbs Beasdale Bank en route from Fort William to Mallaig with BR's West Highland tourist set. *Dr W. A. Sharman*

Above:
The ever-popular *Duchess of Hamilton*, No 46229, heads away from a Garsdale water stop and into the winter sunset on 7 January 1984 with a southbound 'Cumbrian Mountain Express'.
Keith Jackson

support group's hunt round the yard and depot resulted in the 'V2' watering at the diesel depot and the '9F' at the yard control centre. In the meantime, scratch footplate crews had been rustled up to take the engines to Waverley station, where the support coach was shunted to one side for the day. The engines then went on to Haymarket to be coaled and returned to Waverley to pick the train up, which had been worked from London by a Class 47 diesel.

Apart from the two representatives who were to ride on the footplates, the rest of the team was allocated the rear brake van. This was because the train was being filmed and their coach was of a different livery, which was not in keeping with the rest of the train, these circumstance decreeing that this trip was going to be comfortless for the support crew. Matters were not helped when the promised food and drink failed to materialise!

YOU'RE SUPPOSED TO BE FILLING THE TENDER — NOT PUTTING THE FIRE OUT!

The train was pulled up into Greenloaning loop, where the locomotives were watered whilst the passengers were entertained by an arranged parachute drop and the support crew struggling from one end of the train to the other on the railway embankment to ensure that the engines were properly watered and to check that all was well.

After this, they returned to their location at the rear of the train. So on to Gleneagles, where the passengers detrained. It was at this stage that they were invited to the restaurant car to be given food and drink whilst the train proceeded to Perth. Unfortunately, the hot meal, which would have been just the thing after being on the go for 36hr, was not available. But there was dinner to look forward to when they got back to their own coach, or so they thought. The train was taken on empty to Perth with the two engines going to North Yard for

Left:
The 1989 debut to main line running of Class K4 2-6-0 No 3442 *The Great Marquess* was an emotionally-charged event. The locomotive's owner, Lord Lindsey, wanted to see the 2-6-0 running on the West Highland line before he died, which he just managed to do thanks to the efforts of his many friends. Here, No 3442 passes Barrow Hill on 30 June 1989 heading towards Sheffield while working its main line test trip, an operation insisted upon by British Rail before a locomotive can be granted its main line certificate. *Dr W. A. Sharman*

turning. Some of the group returned with the empty stock to Edinburgh whilst others got the service train. The two engines were to return light after taking water to be reunited with the support coach.

But all was not over, as members of the group returned to Edinburgh only to find that their coach had been moved and they were initially unable to locate it. However, one resourceful person, the cook, did locate the coach locked up in the GPO compound and gained access via the platform end to commence the evening meal's preparation.

He was dismayed to find that the others had not joined him by the time the meal was ready because they had been told that they could not join the coach whilst it was in the compound. This could be until the engines arrived back from Perth. By this time it was dark and frosty and they were very hungry indeed. So strong representations by one and all achieved access, much to the cook's relief. They were delighted to find their dinner ready! Eventually the coach was reunited with the engines, only to find that no footplate crews had been rostered again to take the engines back to England. Once more, scratch crews were assembled and, with sighs of relief, the engines headed south into the darkness. Again water was taken at Berwick and Newcastle followed by *Evening Star* departing for the North Yorkshire Moors Railway from Darlington.

Everyone was now looking forward to a quick run home across the racetrack to York. But alas, no. There was a tanker train in front with a problem needing the fire service's attention. So the run home was heavily delayed whilst this incident was cleared up. They arrived back late and had to dispose of the engine and put it away before finally heading home for their beds, with some having work to go to within a few hours.

Fortunately, trips like this are the exception and many good times are had amongst all the hard work. Particularly pleasing is the goodwill engendered by taking the National Collection locomotives out on the line. Not only does it bring pleasure but it encourages the public at large to come to the museum at York to visit the static exhibits.

Before any of the aforementioned strife and glamour of running on BR, comes the supreme challenge of restoring your locomotive to BR specification. This means a statutory major overhaul at least every seven years. It is this achievement that earns today's men of steam their ticket to ride. But that is another story.

Below:
A surprisingly popular locomotive for steam railtours is the 'balanced' '8F' 2-8-0 No 48151. In this photograph taken on 20 May 1989 the '8F' is passing Barrow Wood on the Settle-Carlisle line en route from Manchester to Carlisle.
Dr W. A. Sharman

New Trains for the East Coast main line

Colin Boocock

The official publicity material described it as 'the most powerful train ever to have run in Britain'. Soon afterwards a popular railway magazine reported the said train to have equalled the British Rail speed record. So the new train must be something special. When in fact it is not just one new train but the first of a fleet of 31 trains which is being given these accolades, the public could be expected to sit up and take notice.

The InterCity 225 trains which are now running on the East Coast main line are worthy of the accolades which have been bestowed upon them, as travellers between London King's Cross and Leeds have already found out. Smart looking, very fast, comfortable, quiet and well appointed, the overall ambience is of a unit train designed with the customer in mind, a customer of good taste, demanding high levels of quality and service, in standard class as well as in first.

This is as it should be, for British Rail surely had to 'pull the stops out' to produce a train which could be seen to be much better than the highly popular InterCity 125s, the 'High Speed Trains' or HSTs in common parlance.

Specifying the new East Coast replacement trains was no easy task. The HSTs had been in service on the King's Cross to Edinburgh run since 1979, and had proved so popular that Edinburgh had had to be connected to London with an hourly InterCity train in order to cope with the demand. Services from Newcastle and Leeds and semi-fasts from York were also of necessity normally scheduled on an hourly pattern. With 4,500hp between the two power cars on each train, the HSTs enabled regular 125mph running to be established. Equipped with Mk 3 coaches, they set new standards of ride and quietness, and in internal comfort. The InterCity 125 was a very difficult act to follow.

Above:
The first public train to be hauled regularly by Class 91 locomotives was the 17.36 from King's Cross to Peterborough. The train is seen before departure on 31 March 1989 headed by No 91008. *Dr Iain C. Scotchman*

Why was replacement necessary at all? It is true that the IC125 fleet had only just been refurbished when the 225s came along. After 10 years each set had accumulated around 2 million miles of running (more than twice that attained in their full lives by the best steam locomotives in former years!). Maintenance and timely component renewal could enable them to run on for many years. However, the situation has to be seen in a wider context.

The East Coast main line was a clear candidate for electrification if there was to be any future in the extension of electric traction elsewhere in the UK. Also, IC125s are by their very nature expensive beasts (even though their earning capacity is very high) because they have virtually two complete diesel locomotives in each trainset. This also ties down their range of operation to those places where they can be fuelled and serviced every 24hr. Diesel locomotives are expensive to maintain and overhaul because of their inherent mechanical and electrical complexity. The special design of the HST power cars virtually prohibited them from any duty other than as part of an InterCity 125 set. When a set was idle, so were the two diesel power cars. Modern technology was seen to offer opportunities for development of a train which did not exhibit these disadvantages.

As with all major schemes in the past, displacement of trains from a route which is to receive new stock produces opportunities for elimination of less satisfactory vehicles on other routes. As it happened, the travel boom of the late 1980s reinforced the need for new trains to be made available to the InterCity business. The conditions were right for a scheme which put the wires up from Hitchin (the limit on the main line of previous electrification from King's Cross) to Leeds and eventually all the way through York and Newcastle to Edinburgh.

But could a train be produced which had to be much better than an IC125? Once it had become clear that there was a case for completely new trains, business chiefs at InterCity began to put together an outline specification of the train they wanted. While 125mph was all that the market demanded for the next few years, it was foreseen that faster speeds would be demanded relatively early in the life span of the new trains. Calculations of what was practical showed that a top speed of 140mph would give useful journey time reductions, while enabling InterCity to avoid for the time being the need to buy tilting trains. To reach such high speeds however, certain design requirements had to be met. Track forces needed to be low. Vehicle ride and sway had to be controlled, sway particularly in view of the need to keep the locomotive's pantograph safely tracking the overhead 25kV wire. Weight ought to be kept down, but militating against this was the need to increase the comfort factors for the customers. Air conditioning, sound and heat insulation, and safety features all add to the weight of passenger carriages. In the end, as is always the case when conflicting needs require to be balanced, very careful optimising was necessary.

The result of all this, as everyone reading this already knows, was the development of the Class 91 locomotive and the Mk 4 coaches with their associated driving van trailers. The order for 31 unit trains envisages each train formation being led by a Class 91 at the north end with eight, and later nine, Mk 4 coaches with a streamlined driving van trailer (DVT) at the rear. Southbound the formation would work in push-pull mode with the unmanned locomotive propelling from the rear, receiving its control codes from the driving cab of the DVT through

Below:
On 9 August 1989 the first set of Mk 4 coaches was undergoing testing on the East Coast main line. Headed by Bo-Bo No 91010 it is seen on the down fast line at Peterborough. *Colin Boocock*

Right:
The 'Yorkshire Pullman', 17.50 to Leeds and Harrogate, was formed of an InterCity 225 train set from early October 1989. The driving van trailer is loaded with parcels and mail before departure from King's Cross on 18 October 1989. *Colin Boocock*

encoded signals passed along the train wires.

For one locomotive to replace two diesel power cars and yet retain the advantages of unit train operation forced the designers into some unfamiliar areas. Some 6,000hp were needed to provide enough tractive effort at high speed to sustain 140mph running, as well as to accelerate the designed load of nine coaches plus driving trailer. Yet one locomotive could sensibly only stand on two four-wheeled bogies, half the number used to power the IC125s. A Co-Co layout was of course possible, but why use six axles if four would do the job? Using separately excited traction motors and wheelslip control (not unlike the principles employed in Foster Yeoman's Class 59 freight locomotives) would enable each axle to support very high tractive effort in a way not previously seen on a British Bo-Bo locomotive. Modern electronic technology would enable this to be done.

But would not four large traction motors be too heavy for the track forces which had to be kept low? Clearly, the motors would have to be mounted in or on the locomotive body, not on the bogie frames, if unsprung axle weights were to be minimised. On the Advanced Passenger Train power cars (the most satisfactory of the APT vehicles be it said) the motors had been mounted inside the bodies on the floor, and drove the wheelsets through inclined cardan shafts and gearboxes. The gearboxes took mechanical technology beyond previous railway experience and were never entirely successful. A quite unique solution was developed for the Class 91. The motors were suspended from the underside of the bodyshell, occupying the space which they would have filled had they been located on the bogies. Short, inclined shafts were fitted, driving through gearboxes of sturdy design plus a conventional quill drive to maximise the freedom of movement of the wheelset. Thus the weight of the motors was kept off the bogie's frames and unsprung axle weights were very low for a locomotive of such size. Because it was known (from APT experience) that the gearboxes would work in a punishing environment, two designs were ordered, from Voith and David Brown. In the event, both have performed successfully in laboratory speed/load tests.

Another feature which has led to a lower-than-usual centre of gravity in the Class 91 design has been the mounting of the transformer by suspending it below the locomotive body between the bogies. This helps considerably in controlling sway at speed.

The electric traction equipment of the Class 91 uses a thyristor phase angle control system supplying dc traction motors, similar to that on the BR Class 90 locomotives. The motors are separately excited, which gives benefits for adhesion as mentioned earlier. Somewhat unconventional is the way that the motor fields are fed from the batteries through gate-turn-off thyristor choppers which enable accurate control of rheostatic braking to be achieved. The equipment is rated at almost 7,000hp at its 1hr rating or 6,000hp continuously.

In addition to rheostatic brakes, the locomotives have friction brakes in two forms. There is a disc brake attached to the 'free' end of each traction motor armature shaft. There is also a set of single-sided brake blocks on each wheel tread.

The complete locomotive has a streamlined cab front at one end and

Below:
Simple but effective, the design of the SIG bogies under the Mk 4 carriages is without headstocks to the frames. This view shows one secondary suspension air bag, the large yaw damper to control stability at speed, and the coil primary springs over the axleboxes.
Colin Boocock

Above:
The interior of a Mk 4 standard class saloon illustrates how the layout breaks up the visual continuity by means of the seats being pitched differently on each side of the centre aisle, and by the dark glass screens. Two colours are used in the moquette seat covers, which further aids the illusion. *BR*

is blunt ended at the other. The blunt end would normally be the end which butts up against the first coach of the train, so that aerodynamic disturbance can be minimised. (A streamlined cab against the blunt front end of a carriage is aerodynamically unsatisfactory.) However, because of the need for the Class 91s on occasion to haul southbound trains conventionally, the blunt end is provided with a pleasantly styled cab front. Both the streamlined end and the blunt end have full sets of retractable side buffers and drop-head buck-eye couplings.

That is not the case within the train of Mk 4 carriages, each of which is coupled to its neighbour using tightlock couplers which reduce drawbar shocks. They also reinforce the unit train principle by discouraging splitting of sets for operating reasons.

Each Mk 4 coach is a 23-metre long vehicle of strong steel construction using relatively simple techniques to facilitate speedy delivery. Noise insulation and bogie ride are two features which are intended to be improvements over the Mk 3 design. Bogies are the relatively simple SIG design from Switzerland, balanced for the Mk 4 vehicle body dynamics and performance requirements, and able to meet the very tight specifications relating to 140mph running and lateral and vertical ride. The all-important secondary suspension is based on large air-springs mounted on the bogie frames with supporting rubber springs. There is no swing bolster. The Mk 3 vehicle BT10 bogie family was the last on BR to use that time-honoured means of controlling laterial ride. On the Mk 4 coach lateral flexibility is obtained through the ability of the air and rubber springs to flex in a horizontal plane.

Internally there are numerous new features to improve the comfort and convenience of travellers. Several toilets, for example, are configured to permit their use by people who are confined to wheelchairs, and there are also basic facilities for nursing mothers.

Decor is modern, generally restful, and is likely to impress rather than surprise. Its aim is to provide a restful environment and bright colours have been avoided in order to achieve this. A good feature is the wide gangway between each vehicle which provides much smoother transit for walking passengers than did the older, more traditional Mk 3 gangway. And a real effort has been made to lay out the seats in such a way that all seated customers are able to see clearly out through the nearest side window!

Entrance doors are modern swing-plug doors which open in response to passenger control buttons, and which enable the guard to close all doors safely at the touch of a button before departure.

Dressed overall in BR's smart InterCity livery, a complete InterCity 225 train set looks to be a homogeneous whole, designed as a unit. The rakish front end of the locomotive linked to the sharper angle of the carriage upper sides give the whole train a sleek look which will continue to please the eye for several decades, I believe, just as do its diesel predecessors, the InterCity 125s.

The first sets are in service now between King's Cross and Leeds. The wires are now energised to York, and the catenary is going up on the long stretch to Newcastle and Edinburgh. 1991, the year of this *Year Book*, will see the full consummation of the years of planning and development which have brought electrification to the East Coast main line. That the new services should be blessed with such superb new trains as the InterCity 225s is something of which British Rail InterCity can justifiably be proud.

Odd Engines 2:

Three Small Tank Engines

Right:
Perched on its overhead runway leading from Cork Glanmire Road locomotive depot coal stage to a nearby quay is the 5ft 3in gauge vertical boilered 0-4-0T *Pet*, photographed in 1963. The locomotive gets no marks for aesthetic design even though its frames are said to have come from an old Great Southern & Western Railway four-wheel tender.
Colin Boocock

Below:
Well into its ninth decade, Lewin 0-4-0ST No 18 shunts old chaldron wagons under the coal staithes at Seaham Harbour, Co Durham in 1964. The purpose of this operation appears to have been to enable recovery of loose coal dropped from the wagons on the staithes above.
Colin Boocock

Left:
Like a baby in a man's world, Portuguese Railways 0-4-0WT engine No 004 sits uneasily under a water crane designed for bigger things. The locomotive was the shed pilot at Entroncamento in 1962. *Colin Boocock*

Pet

Pet was an enigma! It never appeared in Ian Allan's *ABC Irish Locomotives*. Yet it survived in occasional use to the end of steam at Cork in Eire.

Pet was a four-wheeled, vertical boiler, 5ft 3in gauge locomotive of indeterminate lineage. It used to haul coal wagons on an overhead structure running from a quayside in Cork to the main railway locomotive depot at Glanmire Road. Its outward decrepitude was enhanced by a home-made cab.

The Lewin

Equally rare, but again a late survivor, was the Stephen Lewin 0-4-0ST which worked under the coal staithes at Seaham Harbour, Co Durham. No 18 was built at Poole in Dorset about 125 years ago. Its survival as Britain's last standard gauge steam locomotive of south coast manufacture was remarkable, as these 1964 photographs demonstrate. Happily its work included moving ancient chaldron wagons for picking up the coals dropped from the staithes. The old wagons were a good match for No 18's austere appearance.

004

Portugal always looked after its steam locomotives, even its humble shed shunters. Cockerill 0-4-0WT engine No 004 was the depot shunting locomotive at Entroncamento. In 1962 Entroncamento was the northern extremity of electrification on the Lisboa-Porto main line and therefore locomotives were changed there.

More than 70 years old, No 004 (and its sisters elsewhere) was short enough to be able to haul a dead 4-6-0 on to the turntable and to be turned coupled together with the other engine. The locomotive's small coupled wheels and its outside double eccentric valve gear emphasised its odd character as it waddled around the depot.

Left:
This close-up of the Lewin 0-4-0ST at Seaham Harbour shows its rugged simplicity, undoubtedly one of the key sources of its survival. *Colin Boocock*

Below:
Portuguese Railways 0-4-0 well tank No 004 poses on the turntable at Entroncamento in summer 1962.
Colin Boocock

David R. Bissett

Aberdeen Memories

Let us go back through the years (some 70-odd of them) to the hallowed place where I spent many happy hours of my young days.

1920

We are standing at the end of a bridge over the railway tracks. In front of us is the Craiginches marshalling yard, with its lines of wagons and vans of all descriptions. Beyond it lies the long, uphill embankment leading to the bridge on the skyline at Bay of Nigg, where the hills at the tail end of the Grampian mountains slope into the cliffs and the North Sea. Through the arch of the distant bridge we can see the sea.

To our right are fields of grain, cattle and sheep. The country road over our bridge climbs up the steep

Right:
A Caledonian Railway 'Dunalastair' 4-4-0 pulls out of Aberdeen station with a train for the West Coast route to London. In the left background a Reid Atlantic piloting a smaller 4-4-0 awaits departure on an East Coast route train. *H. Gordon Tidey*

Below:
The Reid Atlantics of the North British Railway were impressive and beefy machines. Here is No 868 *Aberdonian* in photographic grey. *Ian Allan Library*

Above:
Coming off the Dee Viaduct at Aberdeen is Class V2 2-6-2 No 60972. The train is the up 'West Coast Postal' in its 1960s condition. The photograph was taken in August 1965.
David R. Bissett/Colour-Rail

hillside past the church and hamlet of Nigg. A few farmhouses stand here and there. It is a very peaceful scene.

Looking back the way we came we can see a small foundry (Harpers) and a sawmill. Beyond these are more fields and trees, and in the distance the grim high walls of the local 'jile' (prison).

Looking across the road at the other parapet of the bridge, we see the signalbox, so close that the signalman can converse in a normal voice to any acquaintance passing by. Behind it is a factory making agricultural machines, and farther back on this side road (called the Laundry Road) there is indeed one of these essential establishments.

However, we are not here to admire the surroundings but to pay homage to the London train, or rather to the engine pulling it! The North British Atlantic locomotives were the acme of perfection to me. The Caledonian engines were no doubt more shapely and colourful, but they were just 'wimps' to the 'butch' look of the Atlantics, which got my vote every time.

Today it is a fine day and everything is great — except for a couple of wagons on a siding about 40yd away. They are packed with open topped barrels full of fish offal (heads, tails and guts) bound for the fish meal factory at Cove Bay on the other side of the hills. After some hours on a sunny day waiting to be attached to a goods train the contents of the barrels are giving off an atrocious pong. No wonder the signalbox windows are shut tight. At the far end of the yard a small 'Caley' engine is flitting here and there among the lines of wagons (like a bee in a rose garden) assembling a train, as the regular clashing of buffers and jerking of wagons shows. Let us hope the smelly load will be part of this train! But the shunters aren't daft. They won't take those wagons up to their end of the yard until the train is almost ready to move away.

Suddenly from the signalbox comes 'ding: ding-ding'. After an exchange of code signals on the bell machine come the clash of the signal levers, followed by the 'clump clump' as the signals drop to 'clear'. It won't be long now!

We make guesses as to which of the engines it will be: *Bon-Accord*, *Auld Reekie*, *Bonnie Dundee* or whatever! Our guesses are cut short when from the direction of the city comes a 'whoof — whoof — whoof' which soon becomes a steady bark as the London express's heavy load is started away from the Joint station. As the minutes pass the sound gets steadily louder and looking backwards across the bridge we can see the plume of smoke as the train passes across the river bridge and heads for ours.

There is only one column of smoke, so there aren't two engines on as sometimes happens with a

very heavy load. It is usually a 'Scott' class 4-4-0 in front, *Quentin Durward*, *The Fair Maid*, or even *Wandering Willie* when an extra locomotive is needed. The train is now very close and the ground is beginning to shake beneath us. There is a momentary dulling of the racket and, preceded by a flurry of smoke, the engine explodes from the arch of the bridge: the squat chimney (it always reminded me of the little flat hat worn by Buster Keaton in his film comedies); the massive boiler; thrashing piston rods ('Ah! it's *Bonnie Dundee*'); the two-windowed cab lit by the glare of the open firebox door as the fireman feeds the fire from the tender which is piled high with coal. Then the train passes — 1, 2, 3, 4, 5, 6, 7, 8, 9, 10 coaches and sleeping cars, followed by two six-wheeled vans packed with fish, melting ice dripping below the doors, and finally a small open wagon (a last-minute addition) full of boxes of kippers for the breakfast tables of the South.

On the rear of the wagon hangs a red oval board which states 'train following'. This is for all railwaymen to know that a second section of the express will be coming along shortly. We watch the long train which has just passed wend its way through the yard, pass the other signalbox, and start its climb up the embankment to the distant skyline bridge which shuts its 'eye' as *Bonnie Dundee* reaches it, opening again as the last van passes through. We listen as the train belts along by the cliff tops on the other side of the hill.

Suddenly the 'eye' of the distant bridge blinks briefly, and down the embankment comes a short, apparently black object. It is the Stonehaven local, popularly known as 'The Boozer' on account of the fact that its late evening passengers include quite a few citizens of the 'Granite City' who for private and personal reasons prefer to make whoopee in the hotels and hosteiries of the various villages between Stonehaven and the city. A small blue 'Caley' 0-4-4 tank engine running bunker first is pulling the two-coach train and it slips through the yard and under our bridge, the fireman leaning out for a breather as the last 2 miles home are all downhill and the engine is coasting on its way. With the departure of the local, a silence falls over the scene and we debate whether to wait for the second section of the express which should be coming soon.

However, there are clouds heading towards the sun and a cold wind has sprung up from the Bay of Nigg, *and* those wagons are still rotting away, so we decide to call it a day. It will likely just be a couple of 'Scott' class engines at the head of the train. As we walk off down the road, a breeze from the hills beyond brings to us the faint echoes of *Bonnie Dundee* still battling along the coast on its way south.

It all seems so long ago now.

1960

Having an appointment in the area, and with time to spare, I decide to visit again my childhood Shangri-La. Approaching the bridge, I see that the farmlands have gone from this side. Instead there is a council housing estate stretching from the lineside back to the top of the rising ground on the left side of the railway lines. There are hundreds of houses of varying shapes with bus routes running through them. Harper's foundry has closed down. However, when I reach the bridge the view on the other side is unchanged, and is of open fields and farms, although in the distance beyond the goods yard a tool factory and a concrete pipe works have taken over.

Left:
Peppercorn Class A2 4-6-2 No 60532 *Blue Peter* was photographed by the author on 2 September 1966 leaving Aberdeen with the 13.30 to Glasgow.
David R. Bissett

Above:
Were these the 'unimpressive diesels' referred to by the author? The up 'West Coast Postal' is seen again, this time near Cove Bay headed by a pair of North British Type 2s with No D6125 leading.
David R. Bissett/Colour-Rail

Right:
In diesel days, Class 25 Bo-Bo No 25002 heads south from Aberdeen on 28 March 1974 with an up freight. *Brian Morrison*

I lean on the wire fence looking at the bend in the tracks which leads to the bridge over the river, as the signals are clearly 'off' for a train leaving the city. As I gaze at the scene I become aware of a 'presence'! Looking sideways and then down I see a head poking through the fence, looking up at at me — a small face, about 5 years old. The traces of some recent sorrow show on his cheeks where he has wiped away the tears with his jersey sleeve, smearing the dirt around his eyes. We consider each other for a couple of seconds, and then he speaks: 'Are ye watchin' the bonnie trains, Mister?' My heart goes out to him immediately. Over a gulf of some 40-odd years, it could be me standing there, in my childhood!

Replying in the affirmative, we then become aware of the approaching train. Onward it comes, off the river bridge, a thick, black column of smoke rising above it. As it nears our position I think: 'Bonnie trains indeed!' The engine is beat-up 'V2' class 2-6-2 caked thick with greasy grime and soot, with jets of steam gushing out from unauthorised places. Yet it still has an air of majesty as it bellows its way up to and under our bridge. I find myself counting the vans and wagons as they disappear, just as I did 'back in those childhood days'. There are 34 of them and a guard's brake van. Coming to myself again I look for my companion but he has darted across the road to get a rear view of the departing train. My spare time is now up so I walk away. The next train will probably be an unimpressive diesel, anyway.

1990

Yesterday I went back to my bridge. To my horror I discovered that my bridge had GONE, replaced by another three times wider to accommodate the double carriageway road which sweeps its way up the Nigg brae and over the hill. Along the road roared a continuous stream of motor cars, vans, double-deck buses and articulated trucks. There was even a set of traffic lights at the other end of the new bridge for the benefit of those vehicles turning into the old Laundry Road. Gone, too, was the signalbox, and the tall signal posts, replaced by electric signals like traffic lights. Gone, also, were the fields and farmlands. An industrial estate had taken over and the land was covered with all shapes and sizes of factories, warehouses, garages, offices, stacks of oil pipes — even a bank. The thought struck me that perhaps my young friend of the 1960s might even be working in one of these places.

Half of the marshalling yard where the empty fish vans once stood now held oil tanks, cement silos and assorted sheds and parking spaces. As I stood aghast at the sight before me I became aware of the stink! No, my fish wagons of 70 years back had gone, as had the old fish meal factory at Cove Bay itself. It was a more deadly smell — fumes from the exhausts of the road traffic rumbling past. This was no place to dally if one wanted to stay healthy, so, with a quick 'Harvey Smith' at a string of articulated lorries grinding past, I walked away sadly down the road.

As I walked, I fancied that above the racket and noise of the passing traffic there came yet again the faint echoes from the hills of *Bonnie Dundee* still battling on along by the cliff tops with its heavy train, southward bound!

I DON'T CARE WHAT YOUR PANEL SAYS — I THINK THERE **IS** SOMETHING IN THE SECTION!

Photo Feature: Green Diesels

Early British Railways colours for diesel traction followed the pattern adopted for steam locomotives in most respects. Diesel locomotives were initially uniformly painted in Brunswick green. Some classes which were introduced in later years received added decoration with livery bands in lighter shades of green.

While locomotive-hauled passenger stock was bright carmine and cream, or from 1956 maroon (except on the Southern which retained its own green), diesel multiple-units received Brunswick green lined out in cream. Some DMUs later received a version of the brighter, Southern green.

When dirty, the Brunswick green became very drab indeed, but when clean it was aesthetically quite pleasing.

Above:
Painted in Southern multiple-unit green, a Scottish Region 'Inter-city' six-car set starts out from Edinburgh Waverley station in 1960 on an express working to Glasgow Queen Street. The 350hp diesel electric shunting locomotive on the left is in plain Brunswick green livery.
Colin Boocock

Centre right:
Birmingham Railway Carriage & Wagon (BRCW) Bo-Bo diesel-electric Type '2' No D5305 which today, as No 26005, is based at Eastfield, emerges from Hadley Wood Tunnel. The train is the 1.35pm from King's Cross to Peterborough, photographed on 15 April 1960.
Colin Boocock

Bottom right:
Swindon-built B-B 2,200hp diesel-hydraulic No D809 *Champion* coasts down Bruton Bank on 10 October 1959 with the down 'Cornish Riviera Express'. Although painted maroon in their later years, these locomotives began their working careers in standard Brunswick green. *Colin Boocock*

1M13

Above left:
Brand-new Brush Type 4 Co-Co No D1520 leans to the curve between the Hadley Wood tunnels in 1963 with a down unfitted freight on the East Coast main line. *Colin Boocock*

Left:
Also in 1963, 'Western' class C-C diesel-hydraulic locomotive No D1003 *Western Pioneer* heads past Acocks Green with a Paddington-Birmingham express. *Michael Mensing*

Above:
Leominster branch single unit diesel railcar No 55013 pauses at Bransford Junction, Worcestershire, for single line tablet purposes. The number '2' in the front display panel refers to the train classification (2 = stopping passenger train). *Michael Mensing*

Right:
In a totally different league is 'Deltic' Type '5' Co-Co No D9014 heading south away from Tweedmouth with the Sunday 10.50 Edinburgh-King's Cross on 20 May 1962. *Michael Mensing*

The Railways of New Zealand

Robin Russell

New Zealand is a country where the sun shines from the north, and midsummer is in December; there are claimed to be 20 times as many sheep as people! New Zealand Railways (NZR) runs an efficient freight operation, but the passenger service is meagre by British standards: this reflects the New Zealand population level (just over 3 million) rather than a decadent railway. Modern diesels, high-tech electrics and (very occasionally) restored steam locomotives provide the motive power on New Zealand's 3ft 6in tracks.

Trains currently operate on the lines shown on the map, which omits some branches (particularly in suburban areas). The North Island is the more populated and has greater commercial and industrial activity. The South Island (known humorously as 'The Mainland') has higher mountains, colder weather and a more relaxed life style. Visiting New Zealand is worthwhile at any time of year. (Railway enthusiasts may be interested in the trips organised — in association with To Europe For Steam (TEFS) — by Kiwi Leisureworld Tours Ltd, PO Box 5687, Dunedin, NZ.) In the winter, skiing is popular; in summer, walking, swimming and boating are among the favourite pastimes. Preserved steam lines normally operate during the warmer months. Regardless of season, NZR trains run through generally attractive, often spectacular, scenery; enthusiasts will find all too many curves and gradients on the various lines — almost all of them single track — between the most northerly station at Otira, and Bluff in the far south.

Freight carried between these points is largely what one would expect in a farming community which has also developed industrially. Predominant traffic would include meat and dairy products, fertiliser, grain, wool, wood, wood products such as paper, petroleum and manufactured goods. Certain local traffic will be covered as the seven selected lines are described. These are:

Right:
The sun had already set on 5 June 1989 as streamlined 'J' No 1211 *Gloria* storms out of Taumarunui with an excursion organised by Rangitikei Railfans.
Robin Russell

North Auckland — serves the north of the country, with generally farm-based traffic plus clay and coal; handles suburban passenger trains between Waitakere and Auckland (a significant port and the largest commercial and industrial centre).

North Island Main Trunk (NIMT) — spans the 422 miles between Auckland and Wellington (the nation's capital) and carries all varieties of freight. The branch to Mission Bush serves a large steel mill. Nearly all of the route between Auckland and Hamilton is double track. The NIMT was recently electrified at 25,000V ac between Te Rapa, a suburb of Hamilton, and Palmerston North; associated civil engineering work included deviations, easing of curves and gradients, and daylighting (opening up) of tunnels. There is suburban passenger traffic between Auckland and Papakura, and between Paraparaumu (just north of Paekakariki) and Wellington, the latter route being electrified at 1,500V dc. The NIMT boasts the highest NZR station, Waiouru, which is 2,670ft above sea level.

East Coast Main Trunk (ECMT) — links Hamilton (via the Kaimai Tunnel which opened in 1978 and is the longest in the Southern Hemisphere at 5½ miles) with the port of Tauranga, and thence to Kawerau, site of a large paper mill; there is another at Kinleith. The Thames branch serves an automobile factory.

Marton to New Plymouth — handles most farm-based commodities.

Palmerston North to Gisborne — also carries rural traffic.

South Island Main Trunk (SIMT) — covers the 217 miles between Christchurch and Picton (southern terminal of the train/car/lorry ferry service with Wellington) and the 367 miles between Christchurch and Invercargill. Carries most categories of freight.

Midland — joins east and west coasts; noted for Otira Tunnel, nearly 5⅓ miles at 1 in 33, forming an isolated section of 1,500V dc electrified line (recently investigated for diesel traction, but ventilation problems have yet to be overcome). Predominant Midland line business is carrying — to the port of Lyttelton — the excellent West Coast coal; this produces a hot flame with very few sparks and was always preferred by engine crews in the days of steam.

The steam locomotives of New Zealand have a fascinating history. The first varieties were of British origin; after these came a phase of buying from the USA. Eventually, local manufacture was decided upon, using excellent workshops at Lower Hutt (Wellington), Addington (Christchurch) and Hillside (Dunedin). Certain classes, such as the double Fairlie designs, were not a total success; three Garratt locomotives (with Gresley conjugated valve gear) frequently failed and, when in action, all too often caused broken couplings. Success stories would include the British 'F' class 0-6-0 tanks, built in 1872 and active to this day on preserved railways; the 1878 USA 'K' class 2-4-2 engines were fast and reliable. Designed in 1915, the 'Ab' Pacifics won golden opinions throughout the system. In 1932, the 4-8-4 'K' class appeared, which squeezed a 48sq ft grate and

Below:
The railways of New Zealand.

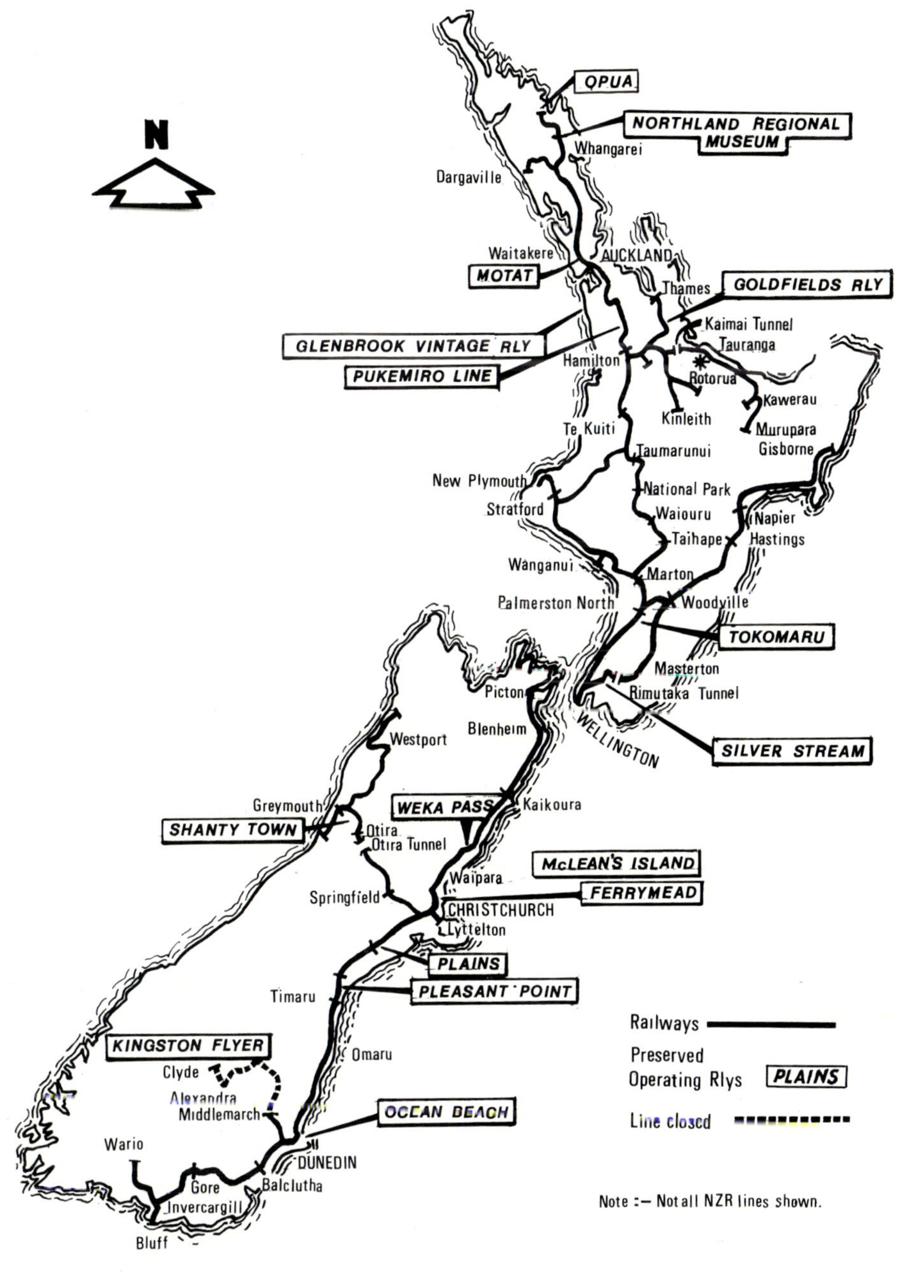

Above:
The 'Silver Fern' diesel set leaves Auckland, bound for Wellington.
Robin Russell

32,740lb tractive effort into a 11ft 6in by 8ft 6in loading gauge, with a 14-ton axle loading. These were followed by 'Ka' and 'Kb' engines, streamlined initially; the latter were booster-equipped and allocated to the Midland line; many 4-8-4s were converted to oil firing. The final significant class was the 'J' 4-8-2, with 39sq ft grate, 26,520lb tractive effort and an axle loading of 11 tons. The 'J' engines were streamlined initially; some were converted to oil firing and designated 'Jb'. The 'Ja' was the final development of the 'J'; some, built by North British, were oil burners (but were not reclassified).

The last New Zealand steam classes incorporated the very best of European and North American design features, including roller bearings, compensated springing, electric lighting, and air operation of fire doors, sanders, cylinder cocks and reverser. Cab layout was excellent, with both injectors on the fireman's side (the left, despite the left line being used on double-track routes). Nevertheless, no damper adjustment was provided, neither was a steam chest pressure gauge nor a cut-off scale on the reverser. Before leaving the steam engines, mention must be made of the Fell 'H' class. These 0-4-2 tanks operated the 1 in 15 Rimutaka incline from 1878 until it was bypassed by a tunnel (nearly 5½ miles long) in 1955. The Fell's two outside cylinders operated in the normal manner, but an inner pair drove horizontal wheels which gripped a special rail, mounted centrally on the sleepers.

The Fell engines were outlived by conventional steam traction, which lasted on the North Island until December 1967, and on the South Island until October 1971. Meanwhile, the first diesel-electric loco-

Left:
The handsome lines of the English Electric diesels are shown in this photograph of Class Dg No 2255 taken at Ferrymead. On the left is 1872-built 'F' class No 13 *Peveril*. *Robin Russell*

motives were the English Electric Bo-Bo 'De' six-cylinder 660hp shunters introduced in 1952. In the event, they proved to be valuable on suburban passenger trains and certain main line work and had long lives. The next English Electric (EE) design was the handsome streamlined, double-ended. 'Df' class; 2-Co-Co-2 locomotives with V12 1,500hp engines.

These were followed by a smaller version, which was single-ended but had a similar cab profile, the A1A-A1A six-cylinder 750hp 'Dg'. Some 'Dg' engines with higher adhesive weight were designated 'Dh'. The 1966 EE Co-Co 'Di' class were too small in number to have the significance which some consider they deserved. It was the 1955 A1A-A1A 'Da' class which had the greatest impact upon steam activity. This General Motors (GM) standard design, with V12 1,425hp engine, initially had a very restricted route availability due to 15-ton axle loading and a generous loading gauge. Track and bridges were soon strengthened, and certain tunnel floors lowered. This work was to pay additional dividends when, subsequently, it meant that containers to ISO dimensions could travel by rail. The 'Da' locomotives (together with a smaller version, the 'Db' class) became preferred to the British designs as being easier to drive and having somewhat lower maintenance costs.

Maintenance was to prove a problem in the early days of the 1972 General Electric (GE) V12 2,750hp Co-Co 'Dx' class; for a while the top controller notch was blanked-off to improve reliability. Careful work by NZR engineers, and rescheduling so the locomotives were given fewer opportunities to cool down between duties, eliminated the troubles and an excellent unit resulted.

Another type with some limitations was the 1968 Mitsubishi V12 1,045hp 'Dj' class. These were a disappointment in so far as the manufacturer's power rating of the engine was optimistic. Nevertheless, the class have already given faithful service throughout 'The Mainland' for over 20 years.

The South Island is now receiving 'Dx' engines cascaded from the NIMT by the new electrics, and has — together with the North Island — had excellent results from the 1979 'Df' locomotives; these are a GM V12, 1,650hp Co-Co design, which has taken the class designation from the erstwhile EE type.

Class letters were again taken from earlier diesels in the case of the 1978 GE 'Dh' Bo-Bo shunter, another numerically small design. A most successful rebuilding resulted in the 'Dc' class, which are 'Da' locomotives with refurbished equipment and new cab outlines. 'Dc' locomotives operate on both islands. The similar 'Dbr' rebuilds of the 'Db' are largely North Island engines. To the observer, there is a somewhat monotonous similarity in the appearance of types such as 'Dbr', 'Dc', 'Df' and — despite the longer hood — 'Dx'. However, the provision of new and upgraded designs shows the outstanding success of NZR in aiming for a modern and efficient locomotive fleet.

Left:
On the Glenbrook Vintage Railway, the preserved 'Ja' 4-8-2 No 1250 *Diana* demonstrates its handsome looks. This locomotive has since run many miles on NZR hauling steam tourist specials.
Robin Russell

Right:
During the author's cab ride on a Class Ef Bo-Bo-Bo electric locomotive, Driver Waretini looks ahead, hand on the power control. *Robin Russell*

Left:
The 'Efs' have some very British features, even though they were designed to meet a specification apparently based on Japanese practice. Visibly British are the yellow cab fronts, the flexicoil secondary suspension, and the 'Brush' cab windscreen layout. These locomotives were photographed at Taumarunui.
Robin Russell

Below left:
Things change in New Zealand just like in the UK! This view of the southbound 'Silver Fern' leaving Hamilton, taken a decade ago, shows a level crossing, signalbox and footbridge, all since removed. A road bridge now crosses the railway here. *Robin Russell*

The efficiency of rail haulage vis-à-vis road was demonstrated by a test run between Auckland and Wellington; the weight of the train and diesel oil consumed were carefully monitored. It proved that — for a given load of freight — up to date lorries use five times as much fuel as a modern diesel-electric locomotive. Additional benefits would be available — in terms of pollution, and importation of fuel oil — in cases where electric traction is concerned.

Considering electric locomotives, the Otira Tunnel is worked by the five 'Eo' class 1,285hp Bo-Bo engines supplied by Toshiba in 1968. However, the predominant interest lies in the recently introduced 'Ef' class electrics. These were built by Brush of Loughborough, a company needing no introduction to British readers. Recent achievements include the BR Class 60s, the motors for the IC125 trains and the production of the '89' class electric. The order for New Zealand comprised 22 Bo-Bo-Bo locomotives, each weighing 106 tons with an axle loading of 17¾ tons (this weight is evidence of the civil engineering work which has taken place on NZR since the days of steam). Rating is just over 4,000hp and tractive effort is 57,500lb. Loading gauge is 12ft 2in by 8ft 10in (again, an increase on the earlier dimensions).

An 'Ef' handles just under 1,000 tons on a 1 in 50 grade, compared with 330 tons for a 'Ka' 4-8-4, 420 tons for a 'Da' and 720 tons for a 'Dx'; furthermore, the electric would climb the bank at a higher speed. There were just a few 'Ef' features where development was required, concerning tuning of the solid state control gear; the regenerative braking system was not totally trouble-free initially. The class is now operating highly effectively and has totally vindicated the decision to electrify.

The opportunity was taken recently to observe the impressive performance of 'Ef' locomotives, by travelling in the cab south from Te Rapa, transferring to a northbound train at National Park. To enable daylight travel, together with return at a convenient hour, both trains had to be goods workings. Express goods exclude four-wheel vehicles from their make-up and run largely in the dark, providing an overnight service between many centres.

The southbound goods was train 601 ex-Westfield, Auckland. Load was 1,185 tons, requiring two locomotives for the climb from Raurimu. Limit is 886 tons for one locomotive; for two, the couplings become the constraint and the maximum is set at 1,280 tons. The engines on this occasion were 30094 and 30105, the former in front with No 2 end facing ahead.

The 'Ef' class is now crewed by only one person, sitting to the right of the cab. To the left of the console are three controls: the reverser, the regenerative brake lever and the (continuously variable) power lever. To the right are the train and locomotive brake levers and the hooter switch. The main dials are for line voltage, current for all six motors, speed and the usual pressure gauges for the air brake. Essential for single-crewing is a two-way radio; this must accompany the driver if the cab is left for any reason. An aircraft-type 'black box' records various parameters including speed. Another innovation is the 'train end monitor', which displays data transmitted from a unit at the tail end of the last vehicle; information includes air pressure in the brake hose and whether the vehicle is moving; also provided is an odometer which can be preset to the train's length and used to confirm that all wagons are safely in a crossing loop.

Loops certainly abound on the single-track route south from Te Rapa. Soon after passing Hamilton at 08.27, train 601 reached 35mph — the cruising speed for a goods train. (Express goods may travel at 50mph; the maximum locomotive speed is 55mph.) Pro-

Right:
Class Dc diesel locomotive No 1569 leaves the eastern portal of the Kaimai Tunnel. The Railway Enthusiasts' Society's excursion on 9 December 1978 was the first train to bring members of the public through the tunnel after its opening on 12 September 1978. The locomotive (which was originally 'Da' No 1508) has since been renumbered Dc 4191. *Robin Russell*

gress was smooth as the train proceeded down the line. The riding of the engine (admittedly over very good permanent way) was excellent, as was the degree of cab silence. This was broken every minute by the sound from the vigilance device; unless immediately cancelled — by actuating the foot pedal, brake, reverser or power controller — it causes the brakes to be applied.

Almost all brake applications were made with the regenerative system, the only occasions being for a few speed restrictions. Te Kuiti was passed at 09.46, after which the climb to Poro-o-tarao Tunnel, mainly at 1 in 70, was effortless. The ensuing descent was taken using regenerative braking, current being fed back into the catenary. It should be mentioned that the overhead wire — energised at 25,000V — is fed from a 50,000V lineside cable, step-down transformers being located every 10 miles or so; there are only four substations, supplying the 50,000V, along the 255 miles between Te Rapa and Palmerston North. Very little traffic between these points is still diesel-hauled, this being local pick-up freights and work trains.

One of the latter was approaching as train 601 headed south, so we waited in the loop at Okahukura (junction for the line to Stratford) until 'Dc' 4375 and some ballast wagons came a stand alongside. After our train had restarted, the short section to Taumarunui was concluded at 11.37 with a stop to change drivers. Ahead lay the steepest part of the NIMT, the world-famous Raurimu spiral, the gradient being 1 in 50 (or 1 in 42 compensated for curve). From Raurimu station the line travels 4¼ miles, including two tunnels and a spiral, to a point which is only 1¼ miles distant from the start in a straight line. The two engines humbled the famous incline, cruising up the hill at a steady 30mph with a current of around 600A in each of the motors; line voltage remained constant at the nominal figure (action must be taken if it drops below 20,000). There was no occasion for actuation of the anti-slip equipment. The 1 in 50 ends at National Park station (on the foothills of the North Island's highest mountain, Ruapehu) where 601 arrived at 12.46. Minutes later, train 700 drew into the loop and crews were exchanged; the northbound locomotives were Nos 30111, 30186 and (running dead) 30140. The journey back was broken for the attachment of additional wagons at Taumarunui, the 1,258-ton train arriving back at Hamilton at 17.47. The day was an impressive demonstration of modern motive power making light work of heavy trains over a mountainous route; NZR is well equipped for the future.

Progress has been notable in other areas, also. The Traffic Monitoring System (TMS) — similar to the British Total Operations Processing System (TOPS) — is an outstandingly effective local product which has been marketed successfully overseas. NZR has also pioneered the 3ft 6in development of Road-Railer vehicles (these are a lorry on a train, or a rail wagon on the road). Signalling on the 25,000V portion of the NIMT uses the very latest technology, achieving total elimination of any interference from the ac supply; communications equipment incorporates the use of fibre optics. Far more simple, but equally effective, is the Track Warrant System, now being introduced to provide a safe working procedure which is cheaper to install, maintain and operate than tablets or CTC and which is suitable for lines on which CTC is not justified. The warrants are raised as a special document; the information may then be transmitted by radio or telephone.

Systems such as these are not the only areas in which efficiency has been improved. Staffing costs — a significant proportion of operating expenses — are being reduced by the introduction of single-person crewing of diesels, in addition to the 'Ef' electrics; eventually, the only exceptions will be on routes with insufficient traffic to justify installation of the necessary radio transmitters. The elimination of guards — and guards' vans — has already been achieved on most goods trains.

Guards have not — of course — been eliminated from passenger services. These comprise, as usual, suburban and main line trains. The former, in the Auckland area, consist of a somewhat limited range of locomotive-hauled carriages; in Wellington there is an extensive timetable of electric multiple-units (EMUs), serving several destinations in addition to Paraparaumu. Apart from a few EE trains dating from the 1930s, the units (fitted with GEC electrical equipment) were delivered by the Hungarian company Ganz-Mavag in 1982.

Some 10 years earlier, two impressive sets of trains (clad in ribbed stainless steel) had been delivered to New Zealand from Japan. The first were locomotive-hauled sleeping cars called the 'Silver Star', which rivalled the South African 'Blue Train' for luxury; alas, travellers on the NIMT proved not to require this level of service and the venture was abandoned. The other were diesel-electric twin railcars called the 'Silver Fern', 900hp units from Nissho-Iwai. These provide a daylight service, again between Auckland and Wellington. (The outstanding success of the 'Fern' to the present day is interesting, as various railcars were used on several routes around the

Left:
'Df' No 6231 drags a 17-coach train round the curve at Newmarket. The train carries Waikato Rugby football supporters en route to Auckland.
Robin Russell

time steam was being phased out, but never achieved total viability.) The other North Island daylight passenger train (apart from the Wellington to Masterton service) is the 'Endeavour', linking Wellington with Hastings and Napier. Overnight, the route of the 'Silver Fern' is plied by the 'Northerner', from which the sleeping cars were recently withdrawn.

The sister train, the 'Southerner', runs during the day on the south line of the SIMT, between Christchurch and Invercargill. Other South Island passenger trains are the 'Tranz-Alpine Express' on the scenic Midland route, and the 'Coastal Pacific' linking Christchurch with Picton, on the north line of the SIMT. The rolling stock for all locomotive-hauled titled trains has recently been refurbished and provides a most satisfactory standard of accommodation (with large windows on the more scenic routes). Main line passenger trains — together with the associated NZR long-distance buses — are promoted as the InterCity service, a familiar title.

Another type of passenger train — all too rare, but very well patronised — is the steam-hauled special. Once regular steam had ceased, a total ban was imposed (as in the UK after the 15-guinea special). However, NZR was tolerant over the prohibition. Firstly, engines to be preserved were allowed to haul excursions to their respective lines. After that, some exceptions were permitted in connection with local activities and the shooting of movies. Perhaps the most significant event was the 1981 Operation Northland, in which a promotion of rail freight service around Whangarei featured the NZR insignia appearing on the front of not only a 'Da' diesel, but also a 4-4-0T obtained from the local cement works.

Four years later came the historic trip, when NZR authorised the double-headed centennial steam train from Auckland to Wellington and return. Motive power consisted of Phil Goldman's 'Ja' class 4-8-2 No 1250 *Diana* (from the Glenbrook Vintage Railway, one of New Zealand's leading preserved lines) and (from Paekakariki) Steam Incorporated's oil-fired 'Ka' class 4-8-4 No 945. Subsequently, these two engines have individually run many main line specials. In 1988 they were joined by Ian Welch's streamlined 'J' class 4-8-2 No 1211 *Gloria*. (The 'J' and 'Ja' — both of which were restored by Ted Pointon, New Zealand's outstanding locomotive engineer — have since double-headed several special trains.) Steam locomotives also operate at the preservation centres indicated on the map; every one of these is well worth visiting.

Invaluable help with this article was received from Tom McGavin and Bob Stott, two of New Zealand's foremost rail authorities. Sincere thanks are also due to Michael Austin, Patrick Duff, Paul Heighton, Richard Norton, John Stichbury and Robert Sweet of the Railway Enthusiasts' Society. Every assistance was provided by NZR management in Wellington and in Hamilton, by Crew Supervisor: Special Duties (ie locomotive inspector) Ken Tenana, by Te Rapa drivers Greg Budd, Dave Simpson and Gerry Waretini and by Taumarunui driver Robert Humphreys. Glenda Arthur diligently edited the text.

Highly recommended further reading includes *NZR Locomotives and Railcars* by McGavin (NZ Railway & Locomotive Society, Wellington); *New Zealand Railways — The First 125 Years* by Leitch and Stott (Heineman Reed, Auckland); *Rail — The Great New Zealand Adventure* by Sinclair (Grantham House, Wellington); and the monthly magazine *Rails* (Southern Press, Wellington).

Addendum
As this book goes to press, information is still coming in of NZR developments including line closures, and changes among the preserved lines. The reader is asked to note that this article is intended to be accurate to around October 1989 – Ed.

Left:
In steam days the 'Ab' Pacifics were favourites with the locomotive crews prior to and after the arrival of the eight-coupled engines. Here, one of the two 'Kingston Flyer' locomotives coasts downhill among the mountains of the South Island. *Robin Russell*

Odd Engines 3:

The 'USA' Tanks

Even to people who lived in Southampton, the 'USA' 0-6-0Ts which shunted the docks were a little odd. They looked foreign to British eyes.

The Southern Railway had put 14 'USA' tanks into service in 1946 to replace London & South Western Railway Class 'B4' 0-4-0Ts which were inadequate for the postwar Southampton Docks traffic. The 'USAs' had formerly been owned by the United States Army Transportation Corps. They were a remnant of several hundreds which had been used throughout Europe to support the Allied drive for victory. The engines' strong haulage ability and their short wheelbase suited them to the sharp curves in the docks. Their extreme ruggedness made maintenance and overhaul simple.

Thus the reader will not be surprised that several 'USA' 0-6-0Ts survived for many years in other countries in Europe. The author has seen them in Austria and France. They have been photographed in Greece. One has been seen shunting Belgrade's main station in Yugoslavia as recently as the mid-1980s.

There are survivors in British preservation in Sussex on the Bluebell and the Kent & East Sussex Railways, and in Yorkshire on the Keighley & Worth Valley line. The Worth Valley Railway calls its example 'Tornado' because of its superbly evocative exhaust sound as it pounds upgrade.

CPB

Below:
This is British Railways 'USA' tank No 30073 as it looked in 1960. The Southern Railway and Region had modified the 'USAs' to suit conditions in Southampton Docks. The locomotives had enlarged bunkers, hinged flaps over the buffers on which the shunter could stand, extra vertical handrails on the tank fronts, vacuum brake, and radio for control around the docks. *Colin Boocock*

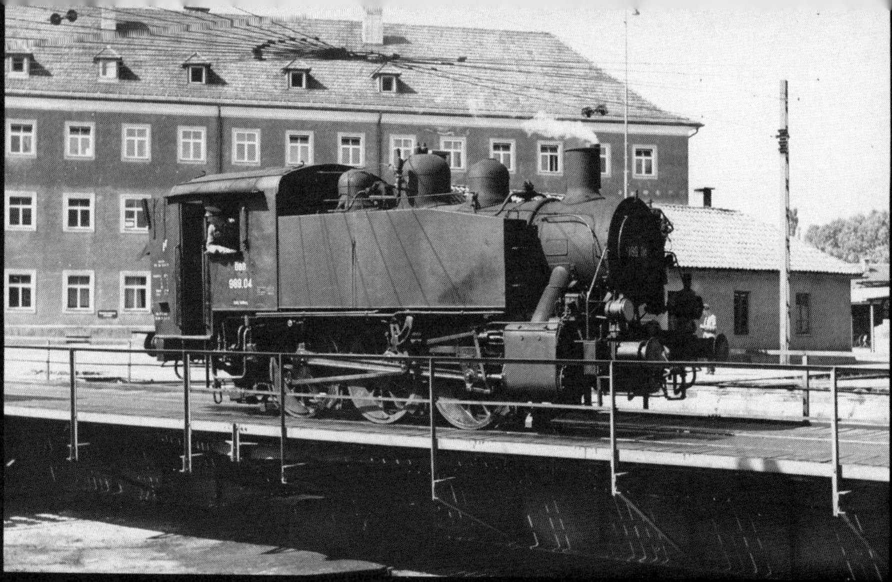

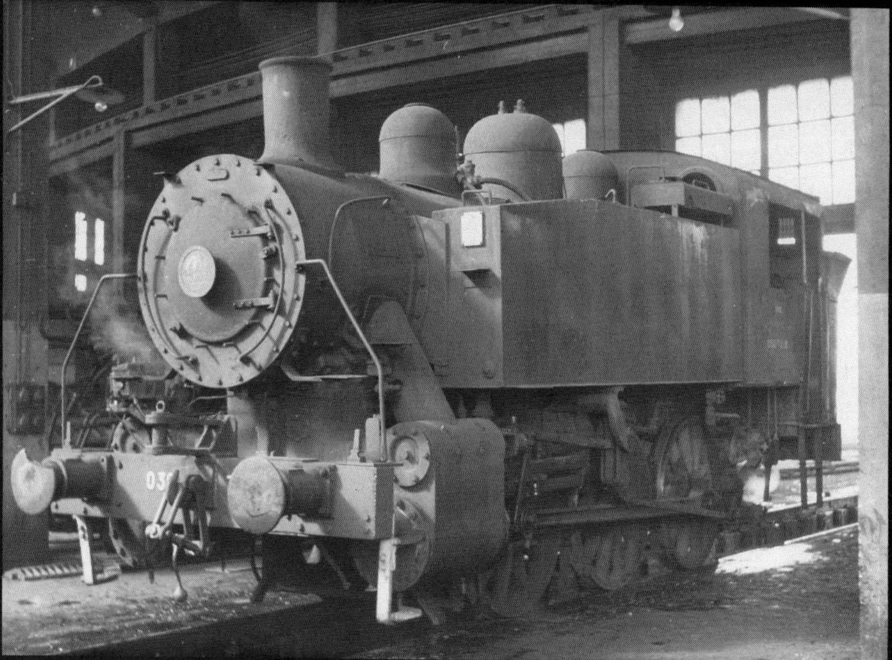

Top left:
No 989.04 of the Austrian Federal Railways is numbered in the 'foreign' series with a class number prefixed by '9'. This one has had its bunker extended upwards, but the cab is original. It is seen shunting the locomotive depot at Salzburg in 1958. *Colin Boocock*

Centre left:
French Railways No 030 TU 18 rests between duties at Lille depot in 1963. This one is relatively unmodified apart from the SNCF plate on the smokebox door, and the chunk out of one of its front buffers! *Colin Boocock*

Bottom left:
'Tornado' earns its nickname! Keighley & Worth Valley Light Railway No 72 starts a train out of Keighley in 1971. The locomotive ran for many years in a most striking and attractive livery based on American practice. Lately it has been repainted in a green approximating to a very light malachite. The KWVLR solved a problem with this class which had eluded BR. Previously these engines had developed hot axleboxes whenever they ran on main lines light for relatively long distances, usually when being delivered to new locations. The Worth Valley engineers fitted a mechanical lubricator to No 72, and the problem disappeared. *Colin Boocock*

Great Western Compound Locomotives

John D. Blyth

The Great Western Railway could not, by any standards, be considered a 'compound' railway. In all its history it owned only five compound locomotives, and those not at all the same time. Of these, two are usually forgotten. Of a later proposal which did not materialise, only a solitary and tentative drawing remains with some sketchy dimensions.

The rest is silence, and even visits by compounds from other railways seem to have been rare, unless by virtue of running powers or on joint sections. We may recall that, when GWR No 5000 *Launceston Castle* was away flattening out the gradients on the West Coast main line to Carlisle, the LMS loaned 4-4-0 No 1047 to the GWR, which was basically a Midland locomotive although built just after the Grouping of 1923. Stationed at Bristol it worked to London, the North and the Southwest. How we would love to know how it performed on the fearsome South Devon banks. Alas, all we have is a record of a run from Paddington to Bath with 210 tons, competent rather than inspired.[1]

Forty years earlier, William Dean had caused two real compounds to be built at Swindon. They were tandem compounds, one for the broad gauge, the other for the standard gauge. Tandem compounds were so rare in Britain that we may recall just what they were. They had pistons for both high and low pressure cylinders one behind the other on the same rod, duplicated, of course, to make a four-cylinder machine. To fit this, one high pressure (hp) and one low pressure (lp) cylinder were arranged on each side of the locomotive in 'line ahead'. It was only in Britain that conservatism called for all cylinders to be between the frames, with a maximum of inaccessibility and ensuring the retention of troublesome crank axles.

Dean however had been upstaged, no doubt by chance, by Matthew Holmes of the North British Railway who, in 1885, had rebuilt his company's 4-4-0 No 224 (notorious as the locomotive which plunged with its train into the River Tay in 1879 and was later recovered) into a tandem compound under Nesbit's patent. Inside cylinders made possible the inclusion of a receiver (the usual outside cylinders resulted in each hp cylinder exhausting directly into its nearby lp counterpart) and means were also provided for the independent adjustment of the high and low pressure cut-offs. The venture was adjudged a success; nevertheless No 224 was restored to 'normal' some years later and was soon afterwards broken up.[2,3] Thus was the British scene set, only the passion for inside cylinders setting it apart from that worldwide.

The Dean Tandem Compounds

If NBR No 224 was a success, the two Dean locomotives, although incorporating much sound and traditional GWR practice, included features which were almost calculated to ensure their failure. Numbered 7 for the standard gauge and 8 for the broad, they were 2-4-0s with the same boiler and a good deal else in common, and they looked remarkably similar. Only the unusually long space in front of the

Below:
The first Great Western compound was No 7 of 1866, the standard gauge tandem. It was quite normal in outside appearance apart from the greater-than-normal length between the carrying wheels and the leading coupled wheels. *LPC*

leading coupled wheels suggested to the observant that there might be something odd within.

There is no convention as to which set of cylinders should be nearest the front of a 'tandem'. Holmes had placed his hp pair in the lead, and Dean did the opposite. On broad gauge No 8 he had good reason, as will be seen. This locomotive, no doubt with the inevitabilities of 1892 well in mind, had been built as a 'convertible', and given success it might have joined its standard gauge sister (had that also been a success) for a long and happy career on the 'slim gauge'. But it was not to be . . .

On the standard gauge No 7 the cylinders for each side were cast together, the two being bolted together along the centreline of the locomotive assembly. The front cover of the hp and the rear cover of the lp cylinder on each side was a single casting, with provision for the insertion of a brass bush to enable the portion of the piston rod which passed between the two to be lubricated. Not only was it found difficult to ensure a proper supply of oil to the rod in this area, but access was also very difficult for dismantling or repair. As a result, lubrication was apt to fail and scoring of the rods was a serious and expensive feature.

The late E. L. Ahrons recorded that, although built for express work, locomotive No 7 never graduated beyond working stopping trains between Swindon and Cardiff. This was almost certainly via Gloucester rather than run the risk of a failure in the Severn Tunnel (which opened for passenger trains shortly after the locomotive was completed). So No 7 would have had to master the stiff climb eastwards up to Sapperton Tunnel, on which

Above:
The broad gauge version, No 8, was not dissimilar, the main visual differences, apart from the width overall, being the line of the valances and coupling rod splasher. *Real Photographs*

the start from Chalford was on 1 in 74, steepening to 1 in 60.

Soon No 7 was laid aside, later to be rebuilt as a 4-4-0 bearing the same number and the name *Charles Saunders*; this was soon changed to *Armstrong*. Rebuilt in 1894, the locomotive lasted until 1928, being renumbered 4171 and receiving a superheated, coned boiler.

If we account No 7, as built, to have been a failure, what must we make of the broad gauge partner in this experiment? Other than the frames, laid down as we have seen with gauge conversion in mind, the layout was similar until we come to the cylinders, which were the cause of most of the trouble. With the two lp cylinders leading, as in No 7, all four were manufactured as a single casting. The lubrication troubles on No 7 did not arise on No 8 because a very novel arrangement of *three* piston rods was adopted. Of these, the central one carried the hp piston only. The other two piston rods passed either side of the hp cylinder from glands placed near the edges of the back cover of the lp cylinders, the rear ends being attached to 'wings' on either side of the crossheads. A similar arrangement for an outside cylinder 0-8-0 type on the Nord Railway of France,[4] built under the direction of du Bousquet, seems to have been all right on such low-speed freight engines, but for one designed for express work it was just not on. Or did the French simply get it right?

Ahrons, who was a Swindon man, had first-hand experience of the temperamental No 8 and has written about it with some feeling. It seems that the early failures were limited (if that is the word) to smashing the pistons, but when a possible period of stability had passed it was decided to try the engine on a Bristol to Swindon train at express speeds. Wisdom dictated that an 'Armstrong' 2-4-0 be coupled in front in case of any trouble — how wise! All seems to have gone well until, with Ahrons on the footplate, they entered Box Tunnel with its rising grade of 1 in 100. Of the next few moments Ahrons has written with restraint in one place,[5] and with deep feeling in another.[6] Suddenly, three out of the four pistons disintegrated, the cylinder casting being smashed. There was a fearful noise and the cab was smothered in steam. In the meantime nuts and bolts and lumps of cast iron showered down on the hapless occupants of the footplate — there was little protection in the form of a cab — and it is indeed fortunate that no severe injury befell Ahrons or the crew. The pilot engine was able to draw the train out of the tunnel.

It is possible that there had been condensation in the lp cylinders, often a problem in early saturated compounds, the relief valves being unable to clear the water. It is also quite likely that the curious double piston rod arrangement may have allowed the lp pistons to shoulder slightly in their cylinders and to wedge solid against the cylinder walls. If so, something had to go — and it did! We do not have to ponder as to why the locomotive was then laid aside for several years. It was eventually rebuilt to conform with the new guise of its standard gauge sister. As No 8 *Gooch*, also a 4-4-0, it had a long and useful life, eventually having a coned, superheated

Left:
This is a rare, but not unknown, photograph of the two rebuilt locomotives Nos 7 and 8, *Armstrong* and *Gooch*, which had previously been the tandem compound 2-4-0s. As 4-4-0 simples they both led useful lives. In this 1910 view at Paddington, No 7, leading, has a No 2 coned boiler while No 8 is still in its 1894 rebuilt condition.
Bucknall Collection/Ian Allan Library

boiler and going finally to the scrap-heap in 1929, by then numbered 4172.

La France, First of the 'Frenchmen'

When the ailing Dean had been succeeded by his brilliant and progressive former assistant George Jackson Churchward no time was lost in up-ending the old traditions. Even before he took over, Churchward's first 4-6-0, No 100 *William Dean*, was on the road. The ineradicable shortcomings which made it forever inferior to its successors were soon eliminated in later locomotives. The next, No 98, is commonly regarded as the precursor of all two-cylinder 4-6-0s built in Britain from then until the end of steam. Those that embodied the maximum of Churchward's principles were those which enjoyed the greatest success. So it is sometimes thought that such a man could not have been the prime mover in the purchase of a locomotive from abroad, and that he must have done so under orders from his directors. G. J. C. was, like all great men, well able to learn from others and there is little doubt that he *was* the moving force in this experiment. Indeed it is known that he was interested in French compounds at least as early as 1899, before his appointment as Locomotive Superintendent and even before the appearance of the first de Glehn Atlantics. In that year he received photographs and drawings of compound 4-4-0s, 4-6-0s and 0-8-0s of the Western Railway of France and almost certainly some details of the Nord locomotives as well.

In October 1903 a French locomotive arrived at Swindon for assembly, mainly in packing cases. This was to be No 102 *La France*, as nearly as possible identical to the 4-4-2s which were daily performing striking work on the difficult lines of the Nord from Paris to Lille and the Channel ports, heavy trains and tight timings being the order of the day. The French equivalent was the 2.643 series,[7] but a few modifications were needed to fit GWR requirements and loading gauge. A shorter, neatly flared chimney (*not* of GWR outline) was fitted, along with modified profiles to the cab sides and roof, wheel treads and flanges to GWR standards, the air brake replaced by vacuum, and the rear-end connections suited to a GWR tender which was not supplied as part of the contract. A smart black livery (allegedly a proposal for a new GWR standard) was chosen. The locomotive was already named on arrival. The builder was the Belfort works of the Société Alsacienne des Constructions Mechaniques, yet No 102 bore a Swindon works number, 2025, and was given a Swindon lot number, 146.

Also from Alsacienne came a real live Frenchman, Conrad Dumâs, to advise the GWR in the assembly, running and maintenance of this engine. Dumâs, as it turned out, liked Britain and the GWR so much that he stayed and was on the Swindon staff for many years. Somewhere along the line he was responsible for the invention of the excellent Instanter coupling.

How wise to enlist the aid of M. Dumâs! And not just to help put No 102 together, for the Nord 4-4-2s were compounds on the de Glehn system, possibly the most complex of all orthodox locomotive types produced and used in quantity.

Imagine the feelings of a GWR top link driver, accustomed to having just a regulator, a reverse and a brake, being confronted with *two* regulators and *two* reverse screws for hp and lp cut-offs, a changeover valve from simple to compound working and vice versa, and a variable blastpipe aperture control! He could work his locomotive five different ways: simple on the hp only; simple on the lp only; simple on all four cyinders; full compound; and by admission of boiler steam to the lp side through a reducing valve, 'reinforced compound'. Indicator diagrams taken on early (and indifferently designed) French engines show good outlines for the high pressure and a nasty squiggle or two for the low pressure cylinders, suggesting that the latter were just sloshing some condensate around![8] From such iniquities did André Chapelon deliver us.

The GWR provided another 4-6-0 to be tried against No 102. This was 171, soon named *Albion*. To make the contest fairer the boiler pressure was increased from 200 to 225lb/sq in and within a year it had been converted to a 4-4-2 as well. No 102 was in fact a very small and light engine, only 64.7 tons in working order and with 33.7 tons for adhesion. The altered boiler made No 171 heavier than 98 with 70.2 tons all up and 36.2 tons for adhesion. In both cases a 4-6-0 would have been better. I have the firm conviction that *no-one* ought to build narrow firebox Atlantics but a 4-6-0 every time. Wide firebox 4-4-2s ought only to be built if there is a very clear need for the wide firebox.

In general running No 102 seems to have been able to do all that was wanted. If there was any slipping on the South Devon banks, it does not seem to have been regular or serious, nor does slipping at starting appear to have been a feature. This was one of the few engines running in Britain which could exceed the not-too-demanding Churchward norm of a 2-ton drawbar pull at 70mph. This is only about 840dbhp (drawbar horse-

Right:
No 102 *La France* passes Old Oak Common West with the inaugural down nonstop 'Cornish Riviera Express' on 1 July 1904. *Ian Allan Library*

power), and without appreciation of the work being done on the Nord it might be that this has caused a recent writer to state that this was 'nothing very special'. Baron G. Vuillet[9] has written giving abundant support to the fact of drawbar pulls being achieved in excess of 3¼ tons at the French permitted maximum speed of 74½mph which is around 1,450dbhp. To be sure, they were driven hard, to the limit in fact, to achieve this, but it *was* done, and it is not criticism of the locomotive to suggest that GWR drivers would not have been willing to handle their engines in a like manner.

It was proved that 102 and 171 could both do all the work required. The GWR 4-4-2 had the slight advantage in maintenance and lubrication costs — which is not surprising — but also in coal and water consumption — which may well surprise. For the moment we will just wonder whether No 98 (4-6-0) or 171 (4-4-2) could have equalled the drawbar pull of the Nord locomotives mentioned above, in those days of saturated steam. Probably not, or if so at great cost for coal and water.

The Two Big 'Frenchmen'

If in some minds the purchase of even one French compound remains an enigma, that of two slightly larger ones must be even more of a puzzle. Surely Nos 102 and 171 had between them proved the home product to be fully adequate for GWR needs? I have never seen it explained why Nos 103 and 104 had to be purchased or what they proved.

However it was, they duly arrived in 1905 from the same works, and built this time to the Paris-Orléans '3001' class 4-4-2 design, heavier at 71.5 tons with 37.2 tons for adhesion and a grate area of 33.5sq ft, not much less than that of the much later 'Kings'. The 'Frenchmen's' cylinders were about an inch more in diameter, too. Again Swindon works numbers 2126/27 were allocated and lot number 157. Nameless at first, they soon had the legends *Président* and *Alliance* on their rear splashers. They were in lined GWR green from the beginning. Like No 102 they had saturated boilers, and slide valves to all four cylinders.

Above:
***La France* ran until 1913 without major modifications. In that year a superheater was fitted to its original boiler as seen here. Note the small steam pipes leading from the smokebox to the high pressure cylinders and the top feed in the place of the original large diameter pipes. The locomotive is seen in GWR green livery.** *BR*

No 102 kept its French boiler for many years, being superheated in 1913 and having a Swindon superheated No 1 boiler fitted in 1916. But in the cases of Nos 103 and 104 an astonishing series of boiler exchanges took place almost at once, involving saturated and superheated No 1 boilers and their own. Indeed at one time each was running with the French boiler off the other! The French boilers never had superheaters. (All the changes have been recorded elsewhere.[10])

Records show that No 171 was compared in dynamometer car trials with all three 'Frenchmen' (Churchward's approval is on record) but the same tale was told of higher fuel, lubrication and maintenance costs on the foreign engines. It may have been hoped that the bigger engines, used at a lower rate of working, might show better figures but this was a false supposition for history shows, time and time again, that a compound shows up at its best when well, if not fully, extended.

Appreciative of the four-cylinder divided drive layout and the smooth ride of the 'Frenchmen', but not convinced over compounding, Churchward produced in 1906 his own four-cylinder 4-4-2 which was numbered 40, and eventually became *North Star*. First of a long line of very successful locomotives, No 40 was rebuilt into a 4-6-0 within three years. Churchward had admitted that its proneness to slipping was more than was acceptable, and it had at least once buckled its side rods in a severe bout of slipping.

Whether or not it was the intention to run No 40 in comparison with Nos 103 and 104 it appears that this was not done and no records are known of No 40 running with the dynamometer car at all. It is remarkable that, in the face of the slipping troubles with No 40, when more two-cylinder locomotives were ordered, 13 were 4-4-2s and only one a 4-6-0. The former were not all converted for some years, and no doubt many a driver had an anxious time on a bad rail as a result.

We do not know much about the 'Frenchmen' in traffic. At first they worked to the West of England, then via Bristol, and No 102 took out the inaugural down 'Cornish Riviera Express' from Paddington on 1 July 1904. Later, all three worked from London on the Birmingham and Worcester lines as well. A pre-1914 *Railway Magazine* carried a photograph of 102 venturing where 103 and 104 could not, on the recently opened through route from the Midlands to Bristol via Stratford-on-Avon and Honeybourne. No suggestions have been made as to where the locomotive was stationed at the time of the photograph, though it was possibly Wolverhampton Stafford Road.

From time to time, some people who possibly ought to have known better have tried to reveal the 'Frenchmen' as failures. Even the late Dr W. A. Tuplin has joined in the fun. A recent suggestion that they did have a serious proneness to slipping has been supported by astonishing claims as to their tractive effort. A figure of some 45,000lbf in simple working has been claimed. Had that writer stopped to think he would have realised that such a figure, which was in excess of a 'King', both types of Gresley 2-8-2, the Lickey banker, and approaching that for the LMS Garratts, could not be right. Plainly he had forgotten the reducing valve, which precluded admission of steam to the lp cylinders at more than about 80lb/sq in. Use this figure and the tractive effort will come down to some 21,000lbf for No 102, and 24,000lbf for 103 and 104. These figures do not give high factors of adhesion, and care would no doubt have been called for in starting. A photograph exists of No 104 leaving Paddington with the steam sanders going full blast, which may tell a story.

No 102 brought the late Cecil J. Allen up from Birmingham in its saturated days, the 87.35 miles being covered in 86min 30sec despite a signal check at the very end of the run. A speed of 86½mph was clocked at Denham, possibly the maximum ever recorded in England with a de Glehn 4-4-2. But the load was only six bogie coaches, 185 tons.[11] Dr Tuplin quotes a run timed by W. H. Randall[12] with a more taxing load of 455 tons in 1913, which reached Leamington in 98½min. Had the engine been superheated, no doubt better could have been done, especially if six coupled wheels had also been provided.

It is one thing to say the 'Frenchmen' were failures, which I do not support, but quite another to suggest that for many reasons they did not suit the particular conditions on the GWR. A locomotive that is a failure is not usually selected to haul the inaugural trip of what was then the longest nonstop run in the world. Nor are failures progressively modified and improved over an appreciable period. Nor are they commonly kept in good main line service for around 20 years.

The 'Frenchmen' were withdrawn from Oxford shed in 1926 to 1928. By this time they were on secondary services and little interest was taken in them. Just at this time another compound proposal was germinating on the GWR.

The Compound 'Castle'

Speculation is rife in this article. We may continue this activity enjoyably when we consider the 1926 proposal to build a compound version of the new 'Castle' class 4-6-0. It is of course well known that the original desire was to build a 'super "Star"' using the No 7 boiler developed in Churchward's final days for the '47xx' 2-8-0s. Although the No 8 'Castle' boiler was an admitted compromise, the new design seemed to be giving satisfaction. That 'Castles' were built in quantity, right into post-nationalisation days, seems to prove their worth. There is however a persistent feeling that they were variable, not only from day to day but from one example to another.

The then general manager was 'tractive effort minded' and gave that dubious figure an importance it has never warranted (and it still will not die!). The LMS was proposing a compound Pacific, so was this an

Below:
Large 'Frenchman' No 103 *Président*, modified with a GWR copper capped chimney which was believed to have been acquired when the draughting was adjusted to improve its original indifferent steaming. *Ian Allan Library*

Above:
No 103 *Président* poses with its GWR superheated boiler which it received in 1914.
Bucknall Collection/Ian Allan Library

influence? Or was it the Southern's first 'Lord Nelson' whose tractive effort beat that of a 'Castle' by a small margin? Again, was there someone in Swindon looking with longing across the water at the performance of the Nord super-Pacifics with heavy trains? And did Conrad Dumâs, still on the staff, say, 'Yes, and you could have something just as good over here!'? It is certainly true that the introduction of the 'Castles' did not give any material advantage over the 'Stars' as regards accelerations, although a small load increase was permissible on some routes.

Only a very sketchy outline drawing remains, on which I have tried here to hang some details and a few dimensions.[13] O. S. Nock tells us a little, and suggests that the compound 'Castle' was W. A. Stanier's brainchild.[14] It had a very GWR look even though the proportions, in retrospect, look a little odd due to all four cylinders being placed virtually in line under the smokebox: gone was the de Glehn layout. The bogie was brought forward by 1ft 10in and its wheelbase increased by 6in to 8ft 0in. The No 8 boiler was retained. In true GWR style, no outside valvegear was to be seen, thus ensuring that the flexibility of the de Glehn system was lost in favour of a fixed cut-off ratio between the high and low pressure systems. A starting valve was provided, so steam was no doubt to be admitted through a reducing valve to the lp cyinders. Had the 'Frenchmen' been copied in this respect, and the pressure again been 80lb/sq in, the tractive effort would have come out at 33,650lbf rather than the 35,500 usually quoted. The latter would be achieved by a curious 91lb/sq in at the reducing valve, while the full 225lb/sq in would have given an impressive but improbable 62,730lbf of tractive effort.

Collett must have known about the compound scheme, yet when Stanier and Hawksworth had worked it up to the stage when Stanier could say that it was time to let Collett see the proposal, it was given very short shrift indeed. O. S. Nock tells the tale: 'They went in to see him, and as Hawksworth put it: "In about five minutes we were out again!"' By the time all this was going on the outline, at least, of the 'Kings' must have been on the drawing board and the command by Sir Felix Pole that the new engine must have a tractive effort of at least 40,000lbf was being designed into it, simply to placate this ambitious but non-technical man. Such a figure could not have been made available in the compound 'Castle' by any sensible means. Thus died the last proposal for a big compound in Britain which had a good chance of success.

Most of us have views for or against compounds. Personally, I am *for*, quite strongly, but not for universal adoption. Yet I am one of the first to accept that the British tradition, no less than its loading gauge, is against it and that its usefulness on our lines, even the Great Western, must have been limited.

John D. Blyth is the Associate Editor of the Journal of the Stephenson Locomotive Society.

Principal Dimensions Of Great Western Compounds

	Tandem Compounds†		102 De Glehn Compound 4-4-2			103 and 104 De Glehn Compound 4-4-2s			Proposed Compound 'Castle'
	7 Std Gauge	8 Broad Gauge	Original Boiler Saturated	Original Boiler Superheated	Swindon No 1 Superheated	Original Boiler Saturated	Swindon No 1 Saturated	Swindon No 1 Superheated	Swindon No 8 Superheated
Cylinders (4):									
Diameter hp	15in	14in	13 3/8in	13 3/8in	13 3/8in	14 3/16in	14 3/16in	14 3/16in	17in
Diameter lp	23in	23in	22 1/16in	20 3/4in	20 3/4in	23 5/8in	23 5/8in	23 5/8in	25in
Piston Stroke:	21in	21in	25 3/16in	25 3/16in	25 3/16in	25 3/16in	25 3/16in	25 3/16in	26in
Boiler:									
Barrel length	11ft 0in	11ft 0in	13ft 7 1/2in	13ft 7 1/2in	14ft 10in	14ft 1 3/8in	14ft 10in	14ft 10in	14ft 10in
Outside diameter									
(Maximum)	4ft 3in	4ft 3in	5ft 0in	5ft 0in	5ft 6in	5ft 1in	5ft 6in	5ft 6in	5ft 9in
(Minimum)	—	—	4ft 9 5/16in	5ft 0in	4ft 10 3/4in	4ft 11 5/16in	4ft 10 3/4in	4ft 10 3/4in	5ft 1 15/16in
Working Pressure (lb/sq/in):	180	180	227	227	225	227	225	225	225
Grate Area (sq ft):	18.3	18.3	29.5	29.5	27.07	33.35	27.07	27.07	29.36
Tubes:									
Fire (No)	241	241	126*	139	176	139*	250	176	201
Outside diameter	1 5/8in	1 5/8in	2 3/4in*	2in	2in	2 3/4in*	2in	2in	2in
Flue (No)	—	—	—	12	14	—	—	14	14
Outside diameter	—	—	—	5 1/8in	5 1/8in	—	—	5 1/8in	5 1/8in
Elements (No)	—	—	—	96	112	—	—	112	84
Outside diameter	—	—	—	7/8in	7/8in	—	—	7/8in	1in
Heating Surfaces sq ft:									
Tubes	1120.3	1120.3	2288.8	1258.2	1686.6	2582.7	1988.6	1686.6	1885.6
Firebox	137.7	137.7	167.0	168.2	154.8	174.3	154.8	154.8	163.7
Superheating Surface sq ft:	—	—	—	280.5	330	—	—	330	262.6
Driving Wheels Diameter:	80 1/2in	80 1/2in	80 1/2in	80 1/2in	80 1/2in	80 1/2in	80 1/2in	80 1/2in	80 1/2in
Weight (tons):									
Engine (full)	44.0	45.3	64.7	64.7	68.4	71.7	70.7	70.7	?
Adhesion	30.9	32.3	33.7	33.7	34.8	37.2	38.0	38.0	?
Tender	?	?	43.1	43.1	43.1	40.0	40.0	40.0	46.7
Total Engine & Tender:	?	?	107.8	107.8	111.5	111.7	110.7	110.0	?
Tractive Effort at 85% Boiler Pressure	12066lb	10041lb	21156lb	19661lb	19661lb	24058lb	23958lb	23958lb	31578lb

*The Serve tubes originally fitted were replaced by 206×2in plain tubes on 102 and 267×2in plain tubes on 103 and 104; the lp cylinders on 102 were reduced to 20 3/4in diameter at about the same time.
†Except for the two tandem compounds tractive effort figures have been calculated for a 4-cyl simple at starting, with 80lb/sq in in the lp cylinders.

Above:
This sketch of the proposed compound 'Castle' is based on the only 'official' drawing that is known, which was published in the RCTS series on *Locomotives of the Great Western Railway*. With its cylinders in line it looks more like a slightly misshapen 'Saint'. Sketch by *John D. Blyth*

Bibliography references

1. Cecil J. Allen: *Locomotive Exchanges*, Ian Allan Ltd, 1949.
2. P. M. Kalla-Bishop: *Tandem Compounds*, published privately, 1949.
3. E. L. Ahrons: *The British Steam Locomotive from 1825 to 1925*, Locomotive Publishing Co, 1927.
4. P. M. Kalla-Bishop: *Op cit* (drawing).
5. E. L. Ahrons: *Op cit.*
6. E. L. Ahrons: *Locomotive and Train Working in the Latter Part of the Nineteenth Century*, Heffer, Cambridge.
7. J. T. van Riemsdijk: 'The Compound Locomotive, Part I', *Proceedings*, The Newcomen Society, 1970-1.
8. André Chapelon: *La Locomotive a Vapeur*, Paris, 1959.
9. Gerard Vuillet: *Railway Reminiscences of Three Continents*, Nelson, 1968.
10. Railway Correspondence & Travel Society: *The Locomotives of the Great Western Railway — Part 8*, RCTS, 1953.
11. Cecil J. Allen: *British Atlantic Locomotives*, Ian Allan Ltd, 1968.
12. W. A. Tuplin: *Great Western Steam*, George Allen & Unwin, 1958.
13. Railway Correspondence & Travel Society: *Op cit — Part 9*, 1962.
14. O. S. Nock: *The Great Western Stars, Castles and Kings, Combined Edition*, David & Charles, 1980.

Photo Feature:

South of England Industrials

Colin Boocock

Most railway enthusiasts forget, or perhaps didn't know, that in the South of England there was a fascinating variety of industrial locomotives to be found in the 1950s and 1960s. There were many Peckett 0-4-0STs, built nearby in Bristol, together with a fair selection of other British common types.

Not everything was steam, either. The photographs on these pages illustrate just a few of the many interesting machines visible in odd corners between Kent and Dorset.

Left:
A year after steam had ended on BR, indeed two years after it finished on the Southern, the Bowater system still linked Sittingbourne with Kemsley Mill and Ridham Dock using narrow gauge steam locomotives. The traffic was wood pulp and finished paper products. In this view, 0-6-2T *Superior* shunts at Kemsley Mill in 1969. *Colin Boocock*

Right:
There was also steam at Hilsea gasworks, Portsmouth. Unfortunately by 19 April 1958 the Barclay 0-4-0ST *Lord Fisher* had been derelict for some years.
Colin Boocock

Below:
Industrial locomotives come in many forms! In the area around the coke ovens at Hilsea gasworks it was necessary to reduce the risk of unplanned ignition, and battery electric tugs were used. This one appears to be about to move a skip loaded with cooling, spent coke.
Colin Boocock

Left:
Northam gasworks in Southampton did not itself use steam in the 1950s, but the nearby coal wharf which fed it did. This is J. R. Wood & Co's 1951-built Peckett 0-4-0ST No 2128 bringing some more wagons in from the BR yard at Northam on 20 September 1955. *Colin Boocock*

Above left:
When motive power needs became desperate, J. R. Wood's successors, Corrall's Wharf purchased redundant 'B4' 0-4-0T No 30096 from BR. They later cut down its cab and gave it nameplates *Corrall Queen*, as seen here in August 1968. *Colin Boocock*

Above:
Deep in the Purbeck Hills of Dorset there used to operate a splendid 2ft 8in gauge railway owned by the clay mining company Pike, Fayle & Co. It had seven locomotives, of which six were steam, which survived to the mid-1950s. One of these was Manning Wardle 0-4-0ST *Quintus* (1854/14), photographed shunting loaded clay wagons on 11 July 1955. *Colin Boocock*

Left:
This was Southampton power station's latest electric locomotive, No 3, built by Greenwood & Batley of Leeds (Works No 1620) in 1939. The power station also had a Baguley locomotive of 1931. By 19 April 1958 both locomotives were out of use. *Colin Boocock*

Eastleigh —

Colin Boocock

A party of young engine spotters visiting Eastleigh works to 'cop' steam locomotives which they would not normally see operating in their native Hampshire would approach the factory up the brick-walled ramp of Campbell Road bridge. From the bridge, which crossed the main Waterloo-Southampton line and the motive power depot entrance at right angles, the enthusiasts would crane their necks to look for unusual locomotives in the works yard before descending the broad cascade of steps to the works gate.

In the 1950s it was policy to take visitors first through the iron foundry, pattern shop and brass foundry, and then through the forge, smith shop, machine shops and fitting shop, before tackling the wheel shop, the boiler shop and finally (and most importantly to the youngsters) the three-and-a-half bays of the locomotive erecting shop. Thus their railway education would be extended beyond mere number taking.

I am sure that I was not the only boy enthusiast to become totally struck by the enormity of the place, by the excitement, dust, smoke and noise we encountered on our tours of the works, and by the sheer detail of the heavy engineering involved. Eastleigh works overhauled all components of our favourite engines, and indeed had manufactured 309 locomotives since it had opened in 1910, culminating in my personal favourites, Mr Bulleid's 'Merchant Navy' class Pacifics. Indeed, from these parties of young engine spotters were recruited many future railway employees who felt they just had to become involved with it all.

Above:
The foundry block at Eastleigh locomotive works on 14 April 1960 looked much as it had always done. The iron foundry occupied the right-hand, longest section, the pattern shop was next, and the brass foundry was the short unit on the farthest end of the block. In the left foreground is a row of steam locomotives awaiting repair. The fitting shop and erecting shop blocks are beyond the left of the picture.
Colin Boocock

Eastleigh works was set up by Mr Dugald Drummond of the London & South Western Railway (LSWR) in 1910 in order to replace and expand the LSWR's locomotive repair and manufacturing capability, which until then had been located in London at the cramped Nine Elms works. The carriage works had moved to Eastleigh a few years earlier.

The layout of the locomotive works at Eastleigh was bold and imaginative for the early years of this century. There were two main factory blocks, the larger one housing the erecting shop in the four high bays on the northeast side. Beyond, across the southeast end, was the boiler shop. In the lower roofed section at the southwest side of the main building were the principal support shops involved in component repair and manufacture. A separate group of buildings against the southwest boundary housed the iron and brass foundries and the heavy forge.

The sights and sounds of the locomotive works dominated the southeast corner of the red brick town of Eastleigh. Actually, the LSWR had built houses for some of its workers adjacent to the site along the Campbell Road. This street formed in effect a long

Main Works for Steam

cul-de-sac, fenced from both the works and the motive power depot by spiked railings. Consequently the road and its houses were known locally as 'Spike Island'.

Anyone living in the area in the 1950s was alerted at 7.07am by the works steam hooter giving its 5min warning of the start of the working day, which itself was marked at 7.12am by another long hooter blast. As the fires were lit in the iron foundry and brass foundry cupolas, and in the forge and smith shop grates, and coke stoves were poked into life throughout the complex, a murky haze soon spread over the area, to add to that already provided by the locomotives in the nearby motive power depot (MPD). Inside the fitting shop on a frozen winter's morning, a cold spanner would stick to an unsuspecting hand until the stoves began to exert their feeble influence. Boilersuit overalls were inadequate protection from the cold of a sharp, icy morning.

The ancient coke stoves had other practical uses! About 10min before the mid-morning tea break someone would place a section of a brass eccentric block (discarded from an overhauled steam locomotive) into the burning coke. When red hot, the block would be retrieved with long tongs and plunged into a bucket of water. The spitting steam having subsided, the gang was then able to wash before tackling sandwiches

Above left:
Newly rebuilt 'Battle of Britain' 4-6-2 No 34082 *615 Squadron* swings over the heads of workers in the erecting shop on 14 April 1960. It is being positioned in readiness to be lowered on to its coupled wheels. *Colin Boocock*

Left:
Looking along the erecting shop on the same day one can see the lines of steam locomotives undergoing overhaul or repair. Classes seen are rebuilt and unrebuilt 'Merchant Navy' 4-6-2s, an 'S15' 4-6-0, a rebuilt 'Battle of Britain', and a BR Standard 2-6-0. One of the tall coke stoves used for heating the shop is in the extreme right foreground.
Colin Boocock

Right:
In the iron foundry a mould on the continuous casting plant is filled with molten iron, and springs a small leak which is about to be plugged. In the foreground are moulds which are cooling after filling. They will be emptied of sand and the new castings removed for fettling on the opposite side of the plant. *Colin Boocock*

(for some, the mid-morning break was always called 'breakfast'), to be washed down by tea, tasting of bromide, from the tea urn. The urn had been delivered to the work location on the flat back of one of the works Lister trucks.

These Lister trucks were part of the local atmosphere, in more ways than one. They were three-wheeled, articulated contraptions with a horrible two-cylinder petrol engine

Below:
Layout and functions of shops at Eastleigh locomotive works in 1961.

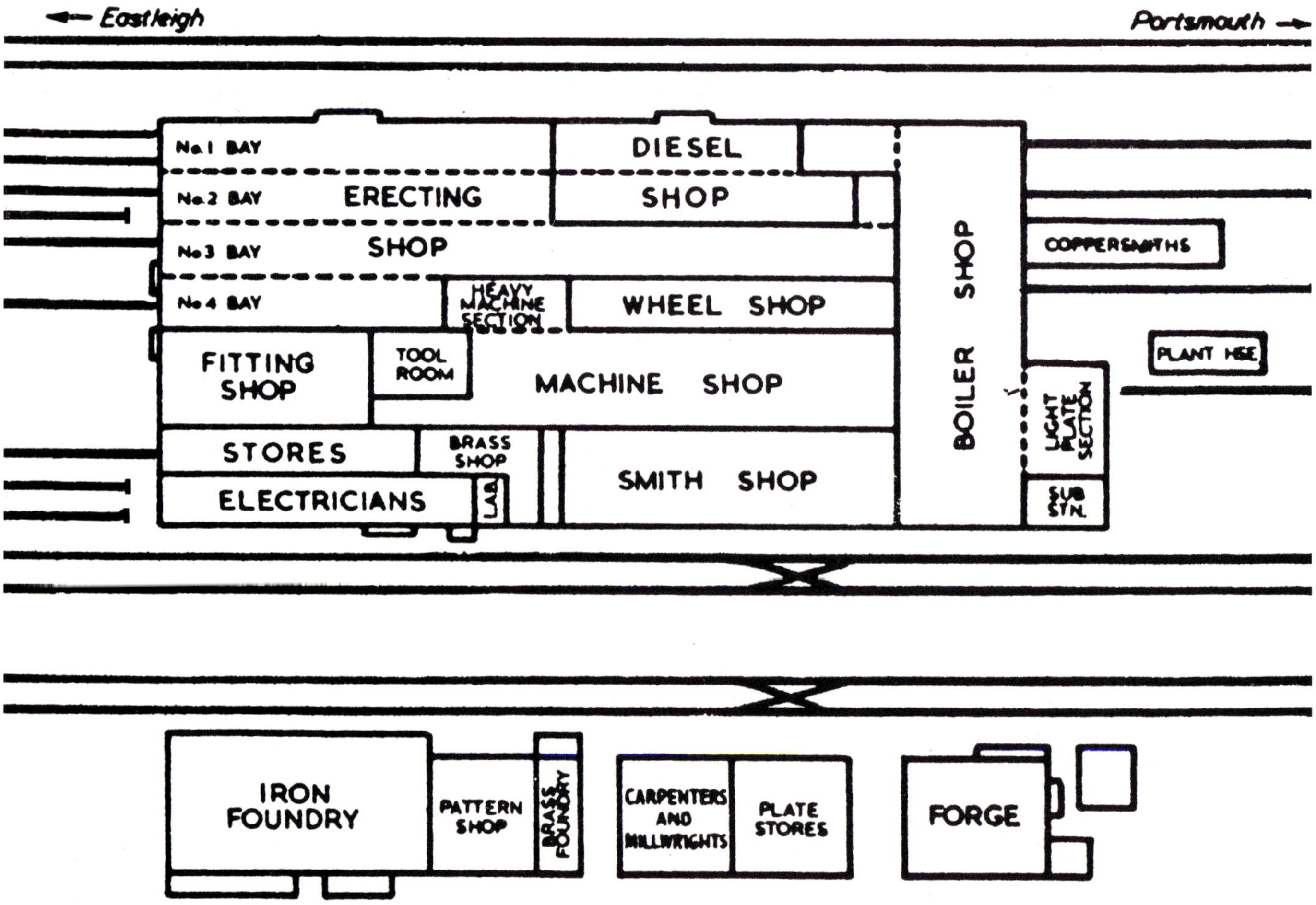

Right:
The biggest castings undertaken at Eastleigh were locomotive cylinders, some of which, like the 'Z' class were formed as a pair of cylinders in a single casting. Here the mould had been let down into a pit to facilitate access for pouring. For such large castings the sand moulds were baked hard in the large ovens seen in the background.
Colin Boocock

mounted on the front. They clattered and stank as they hurtled along the aisles between benches and machines. The trucks would never have survived into the age of the Health & Safety of People at Work Act and indeed they were replaced by quiet, environmentally clean Brush electric battery trucks in later, more enlightened times.

Neither would much of the works machinery have met the requirements of the Act in those days. Groups of lathes and shaping machines were still driven by open, flat belts off rotating shafts, which themselves were driven by belts from large electric motors. The belts, and indeed the machine tools, were relatively unguarded. People just kept clear to survive. The oldest machine was the nut tapper, a crude device dating from 1904 and from Nine Elms operated by first-year apprentices. Most equipment was either original (ie 1910) or at best prewar.

There was some good production equipment in the iron foundry in the form of the continuous casting plant. This was based on a segmented conveyor in a continuous oval on which sand moulds were placed, to be filled with molten iron while the moulds traversed one side of the layout. The castings cooled off as they were carried round one end of the layout, to be emptied of their cast brake blocks or other small components. Sand mixing was automated, as was the creation of the moulds using sand vibrators and presses. By contrast, the production of one-off castings such as locomotive cylinders or smokebox numberplates was very much a skilled, 'hand made' operation using wood patterns.

The works in the 1950s provided employment for 2,500 people of all ages from 15 to 73. I do remember a splendid old fellow in the fitting shop who worked on the connecting rod bench. This lean, lithe, little man used to fit the small and big end bearings. On the heavily built Urie and Maunsell engines this demanded strength to handle the rods, bearings, straps and cotters. He had not retired (and no one insisted that he should, even at 73) because his aim in life was to earn enough money each year to go for a fortnight's binge in Austria!

Life was heavy also in the boiler shop. Indeed, deafness became the stock-in-trade of many boilermakers. In this shop steel plates were flame cut and rolled, and rough edges were cleaned up using pneumatic chisels. Boilers were assembled or repaired with overlapping plates, by drilling rows of holes along the seams and riveting the plates together. The hammer of the rivet guns were supplemented by the hammer of caulking guns which were used to blend the edges of plates, rivets and firebox stay ends to form watertight seals. The worst place for anyone was to be inside a copper or steel firebox with a rivet or caulking gun. Ear defenders would be compulsory wear nowadays.

By contrast the erecting shop was quieter even though it was the centre of some of the most hectic activity. At the top end of the shop the stripping gang worked to dis-

Right:
The wheel balancing machine in the wheel shop was simple but effective. The wheels were spun with the journals held in spring-mounted bearings. Judged from the amount of out-of-balance movement, a calculation was made of the amount of lead necessary to be run into the balance weight sockets of these Bulleid wheels, as marked in chalk on the pair shown here. *Colin Boocock*

mantle an incoming engine which had run 50,000 to 65,000 miles since its last overhaul. This was very dirty work, and the gang normally wore overalls which were impregnated, almost shiny, with dirt and grease. Overhead cranes would lift the locomotive off its wheels and place it on stands. At operating intervals of around five years the boiler would be lifted off, too, for general overhaul.

Boiler work on an intermediate overhaul included the thankless task of clearing and cleaning the boiler tubes. This was the smokebox gang's unenviable task, using screws and brushes on the end of long rods poked progressively into each tube. We did criticise the depots which let tubes get totally blocked. Their specified job of keeping tubes clean, regularly, was infinitely easier than ours was in having to unblock them, and blocked tubes would adversely affect steaming.

A frequent sight in the erecting shop was someone cutting out a crack in a locomotive plate frame, often a 'King Arthur', and having it welded up; even complete frame sections would be cut out and replaced by new. Cylinders were bored out *in situ,* new piston rings sized up inside the cylinders and fitted to the pistons, and gradually the locomotive's parts were reassembled.

In a side bay the tenders were descaled, and rusted plates were replaced by new. Overhauled, newly turned wheelsets were delivered from the wheel shop and whitemetal bearings scraped to fit the axle journals.

At Eastleigh the painters did their job in the erecting shop among everyone else, thus saving downtime of the locomotive which did not have to spend subsequent days in a separate paint shop when it could otherwise be used in traffic.

After re-wheeling, final assembly of valve gear took place, the locomotive was lifted on to rollers at the valve-setting pit, and the skilled work began of setting valves to open and cut off correctly, key work to the ongoing performance of the engine.

A steam test took place in the works yard, and the bigger engines did a light run to Botley (on the Fareham line) and back. Then, at the end of the day a tank engine such as an 'M7' 0-4-4T or the 0-6-2T *Earl of Mount Edgcumbe* would collect all the repaired and overhauled locomotives, all cooled off by this time, and shunt them to the MPD.

The locomotive works in those days was very much a working community with good morale. There was pressure on the often harassed foremen to keep up the pace of work, and sometimes less was done when they weren't looking, but the pace of work was generally kept up by the need of the men to earn sufficient piecework bonus to maintain their earnings. Apart from the Works Manager and his assistant, there was no complex organisation, no planning department and no work study. Piecework prices were negotiated between the men and the shop foreman and, if he was a weak negotiator, men either earned too much or productivity was low.

There was also much friendly rivalry between men of differing backgrounds. In addition to local Eastleigh or Hampshire folk, in the 1950s I met men there who were clearly of London or even Cockney origin, being descended from fathers who had been transferred when the Nine Elms works was closed. There were also several people who had come from Ashford and Brighton during periods of manpower reduction in those places, and who were, perhaps in jest, regarded rather askance by genuine Eastleigh men! All in all, however, the Southern Region was well served by the men of Eastleigh, the main works for steam.

Above:
The boiler shop consisted of two bays like this one, both of them very noisy from the hammering of rivet and caulking guns. All tasks from marking out, cutting and rolling plates to welding of steel fireboxes and flue tube expanding were carried out here.
Colin Boocock

Odd Engines 4:

'Sik Goats'

Visitors to Holland will have been surprised at the sight of tiny diesel shunting locomotives with tall chimneys and canopied cabs. Most large stations and yards have at least one.

These are the '200' class 0-4-0 machines, 85hp diesel-electrics which have simple, basic controls and can be worked by local station staff. Weighing 21 tons, each 'Sik' has a top speed of 37.5mph. The class of around 370 examples was built by Werkspoor between 1934 and 1951. The Dutch nickname for one of these odd but pretty (?) contraptions is 'Sik' or, in English, 'Goat'.

Unlike the other 'Odd Engines' in this book, the 'Siks' are very much in use today and clearly still have a useful future.

CPB

Above right:
In its early condition, but now preserved at the railway museum at Utrecht, is 'Sik' diesel shunter No 311, seen there on 14 July 1989. Note the deep stepboards, the many handrails and the outside controls which enabled the locomotive to be driven by local station crews. No 311 is painted dark green. *Colin Boocock*

Right:
On 4 July 1980, 'Sik' No 320 shunts postal wagons at Deventer.
Colin Boocock

Left:
No 350 appears to be squashed at Venlo! This little 'goat' has brought a Deutsche Bundesbahn Class 141 Bo-Bo under the 15kV wires on 6 September 1985 while it prepares to remove Dutch Bo+Bo No 1160 off an Amsterdam-Cologne express, ready to propel the latter back under its 1,500V wires. There must be an easier way of doing this. *Colin Boocock*

The Veteran Minis

Peter Clowes

Britain's two oldest seaside railways are steaming back

The driver, a grey-haired man in blue overalls and a greasy cap, grinned at me from his padded seat on the tender of a tiny green locomotive. 'Come and have a look', he said.

Two dusty black boots seemed to cover the small footplate, the driver's knees almost brushed the cab's curved steel roof, bright orange flames gleamed and flickered beyond a saucer-sized fire door, scalding hot water dripped from a leaking steam pipe and hazy blue smoke drifting back from a squat black chimney made me cough and splutter.

It was 1935 and I was 8 years old. I was determined to be an engine driver and the 'toy' railway at the Lancashire seaside resort of Southport was the first miniature line I had ever seen.

Another little railway that I occasionally visited in my boyhood was at Rhyl in North Wales. The odd thing is that this line and the Southport one are still in use after nearly 80 years. The tracks, both laid down in 1911, are undoubtedly the oldest surviving ones in Britain.

At Southport two beautiful Atlantic locomotives, their names *King George* and *Princess Elizabeth*, puffed and whistled along a single track bordering the Marine Lake and left me goggle-eyed. I clung to the wooden railings at Pleasureland station and watched them hour after hour. I pestered my parents for ride after ride in the narrow open carriages. Whenever I was missed from the family bivouac of deck-chairs and sand castles on the beach they knew where to find me — glued to those dusty white railings where the little trains rattled to and fro.

The Lakeside Miniature Railway reached its peak of popularity in the 1930s. Pleasureland station had a main platform served by two lines which ended in buffers next to the booking hall. There were three extra sidings for spare green-and-cream toast-rack bogie coaches. A drum-type water tower, a dovecote on a pole complete with cooing fantails, a cage of budgerigars, two snow-white Angora rabbits that sometimes ran among the feet of passengers, a signal gantry with semaphores permanently drooped to 'all-clear' and an attractive slope adorned with rocks and Alpine plants were also among the station's features.

The line, of 15in gauge, ran as straight as a die between floral cuttings, past a rustic engine shed half hidden in a colourful shrubbery, under an Italian-style bridge used by pedestrians crossing a lake from the Promenade to the bathing pool and terminated in the shadow of Southport's long, antiquated Victorian pier. This single-track station was first known as Prince's Park (later renamed Marine Drive). On its wood-slatted platform stood a tank of goldfish. The line was fenced the entire way and there were many footpaths, lawns and benches from which the trains could be seen.

A former postman, Mr George Llewellyn, went into partnership with a local dentist named Ladmore and three others to raise the money to build the Southport track. He leased a quarter-mile-long strip of land along the west side of the Marine Lake and commissioned Miniature Railways of Great Britain Ltd to build a single-track line and equip it with a locomotive and a set of coaches, all built by Bassett-Lowke of Northampton. The designer Henry Greenly travelled to Lancashire to give expert engineering advice.

While the trackbed was being prepared in a shallow cutting that protected the railway from sand-laden winds the contracting company was reconstituted and became Narrow Gauge Railways Ltd, with Mr W. J. Bassett-Lowke remaining as managing director.

The buildings at the line's two simple stations were rustic structures and there was a short and rather ugly tunnel. The line's rake of open bogie coaches was fitted with reversible seats, and four roofed coaches had high glass screens at each end to protect passengers sitting near the engine from smoke and steam. No run-round loops were provided, the locomotive hauling its train from Pleasureland (at first known as White City) to Marine Drive, then propelling the coaches back along the line. Adults were charged 3d, children 1½d.

The locomotive delivered to Mr Llewellyn was the seventh of Bassett-Lowke's Class 10 'Little Giants' to be built at Northampton. It was named *George the Fifth* and measured 14ft 1½in long over buffers. The designer, Henry Greenly, was a great admirer of the Atlantics recently built for the Great Northern Railway and he considered the 4-4-2 wheel arrangement ideal for 15in-gauge railways — the firebox could be made much wider than a 4-6-0 design would

Below:
King George V **looks immaculate in this 1935 photograph taken at Southport. This is probably the locomotive built as *Prince Edward of Wales* in 1912, Bassett-Lowke No 20, which was rebuilt and renamed in 1934.**
A. Phillips/Real Photographs

allow and coupled driving wheels set close together could negotiate the quite sharp curves that lines in amusement parks often required. Speeds up to 30mph were attained and loads of 2½ tons hauled with ease during tests with the first 'Little Giant'.

George the Fifth (the present Queen's grandfather had just become King) had 18in diameter driving wheels, two 6in-stroke cylinders, a 5ft boiler with 37 tubes and a working pressure of 120lb/sq in. Although it steamed well the demands made upon it during Bank Holidays and the Lancashire Wakes weeks proved to George Llewellyn that a more powerful locomotive was required for use in a popular amusement park. Henry Greenly had already reached the same conclusion. Twelve months after the Southport line opened an 'Improved Little Giant' arrived from Bassett-Lowke's workshop and *George the Fifth* was sold to the new line at Rhyl.

Prince Edward of Wales, only the second Class 20 engine to be built, had a slightly bigger boiler and cylinders, a grid superheater and a working pressure of 125lb/sq in. There were straight splashers over the driving wheels, forced lubrication for the cylinders and outside bearings on the pony truck. The locomotive was now 14ft 9in long and weighed a ton more than its Class 10 predecessor.

The number of passengers wishing to travel along Llewellyn's short length of seaside line continued to grow and in 1919 an 0-6-0 side tank, *Katie* was purchased from the Ravenglass and Eskdale Railway to assist *Prince Edward of Wales*. The Cumberland line was glad to get rid of this 23-year-old veteran. It had a tall chimney, a big brass steam dome and a small marine-type boiler that failed to develop enough steam for the line's steep gradients.

But Llewellyn, too, found it unreliable for the regular two-trains-in-steam service he now required and three years later he exchanged it for another Class 20 'Improved Little Giant' from the Fairbourne Railway in Wales. In the meantime, however, Llewellyn had bought another 'Little Giant', *Mighty Atom*, from the Sutton Park Railway in Birmingham. He rebuilt it with continuous splashers and gave it the name *Prince of Wales* before selling it to a Great Yarmouth line in 1929.

In 1931 a disastrous fire swept Southport's wooden engine shed and the line's two steam locomotives were seriously damaged. The company went into liquidation and in 1934 was bought from the Official Receiver by Harry Barlow, a talented local engineer who had been employed by the railway since 1928.

Mr Barlow set about resurrecting the line. He redesigned Pleasureland station, removed the tunnel, reconstructed the engine shed, workshop and overbridge and successfully rebuilt and renamed both locomotives — *Prince Edward of Wales* becoming *King George* and the Fairbourne engine *Princess Elizabeth*.

Particular attention was paid to the boilers. 'You must get them watertight. It's essential,' he once told me. Boiler maker Bernard Cook had them both tested to 160lb/sq in. Mr Barlow's reputation as a miniature locomotive engineer grew. In his Upper Aughton Road workshops over the following years he and his team produced or rebuilt 20 more engines, many of them diesels, for his own line and other 15in-gauge railways. Three of the most interesting were diesel-electrics built to the bizarre designs of cartoonist Rowland Emmett for the Festival of Britain at Battersea Park in 1951. Others were built for pleasure lines at Saltburn, St Annes and for the track along Southport Pier.

In 1946 Mr Barlow was able to lease a further piece of land beyond Southport Pier. He built a concrete bridge over the line, swung the rails through a 90° curve into a new Marine Drive station with three neat stone-flagged platforms and linked the resort's main pleasure beach with a popular amusement park on the sands known as Peter Pan's Playground (now Happiland). On one day 13,000 passengers were carried on the railway.

Mr Barlow retired to his home in Southport in 1968. The syndicate that bought the Lakeside Miniature Railway turned their backs on steam, however, and switched to three Barlow-built streamlined diesels.

Princess Elizabeth was sold to Mark Bamford, a Staffordshire collector, and since 1971 has been running on a half-mile length of line at the Carnforth steam centre in north Lancashire. Early in 1989 it underwent a major overhaul, being stripped down in Steamtown's narrow-gauge workshop, its wheels retyred and its broken axle springs replaced. *King George* was sold at Christie's for £1,000 in 1969 and has been hauling trains in its green Southport livery at Whorlton Lido, Co Durham, for the past 13 years.

Smart streamlined diesels still haul trains along the Lakeside Miniature Railway but it is good to know that steam is making a comeback. A new boiler, running gear and tender for a 15in-gauge steam locomotive are already in the workshop and parts from the 1909 'Little Giant' *Red Dragon* (which ran at Margate as *Prince Edward of Wales* from 1920 until cannibalised in 1968) have been purchased by the Southport line's present owner Mr John Spencer. Enthusiasts are

Right:
A photograph taken around 1914 at Rhyl during what appears to be a formal visit to the railway. Bassett-Lowke 'Little Giant' 4-4-2 *Prince Edward of Wales* is seen pulling away with a train.
Peter Clowes Collection

looking forward to hearing a piercing steam whistle sound across the choppy waters of the Marine Lake once more.

Fifty miles from Southport lies Rhyl, where steam, happily, is still in use, but only just, on the resort's equally-historic miniature railway.

The Bassett-Lowke company, Miniature Railways Ltd, was attracted to the Welsh seaside town in 1910 when it learned that the local council had reclaimed ground at the mouth of the River Clwyd and built an amusement park around a 31-acre excavated lake. Henry Greenly surveyed the site in December of that year and described it as 'ideal' for a miniature railway operating on the principle of a single track with one train running at a time.

Construction work started in 1911, standard industrial 12lb-to-the-yard flat-bottomed rails being laid on steel sleepers supplemented with wooden ones sawn from full-size railway sleepers. Shingle from Rhyl beach was used as ballast.

The track — its steepest gradient was 1 in 160 and it was a mile long — ran around the Marine Lake. The single station's wooden booking office was surmounted by a handsome clock tower. A gantry carrying starting and shunting signals straddled the line to a 100gal water tower. An engine shed, also of wood, stood on a siding at the opposite side of the lake.

It was on 1 May 1911 that the line was opened to the public. *Red Dragon*, the Bassett-Lowke 'Little Giant' that later ended its days in Margate, was purchased from an exhibition line at the White City in London and appropriately renamed *Prince Edward of Wales* — the King's eldest son was to be instituted Prince of Wales at Caernarvon Castle the following month and the principality was ablaze with patriotic flags. The engine was painted in Caledonian blue and hauled a train of six four-wheel chocolate-and-cream open coaches, each carrying eight people.

Holidaymakers flocked to the line. Earl Grosvenor, son of the Duke of Westminster, asked to travel on the tender and this extraordinary practice proved so popular that the public were charged a shilling (children 6d) for a 'footplate pass'. How room was found for the driver, Fred Smithies, and the railway dog Kelly that always somehow managed to perch on the tender remains a mystery to this day.

On August Bank Holiday Monday that year the little engine made no fewer than 93 journeys around the lake, covering nearly 100 miles and carrying 5,000 passengers. It is little wonder that when Narrow Gauge Railways Ltd sold the line to the Rhyl amusement park company in 1912 it was decided to introduce an additional locomotive.

This was the 'Little Giant' *George the Fifth* from Southport. A loop line was laid in the station, a new two-road engine shed complete with inspection pit built and a tunnel created under a roller-coaster.

The manager was Mr Albert Barnes, a strong-minded, efficient man instantly recognisable by the Stetson hat he invariably wore. He called in Henry Greenly to design a new type of locomotive bigger and more powerful than the earlier 'Little Giants'. World War 1 delayed matters, but during the 1920s six of the new engines were built by A. Barnes & Co, the manager's family business, at its Albion Works in Rhyl.

The first Greenly-designed 4-4-2 locomotive, No 101 *Joan*, which entered service in 1920 when *Prince*

Above right:
This photograph of the Barnes Atlantic No 105 *Michael* was used for the abortive 1986 auction. The caption describes the locomotive as being painted in royal maroon with polished brass fittings including the original makers' nameplates. *Ian Allan Library*

Right:
Photographed in April 1989, the 1911-built Bassett-Lowke 4-4-2 No 18 *King George V* is now working at Steamtown, Carnforth. *Peter Clowes*

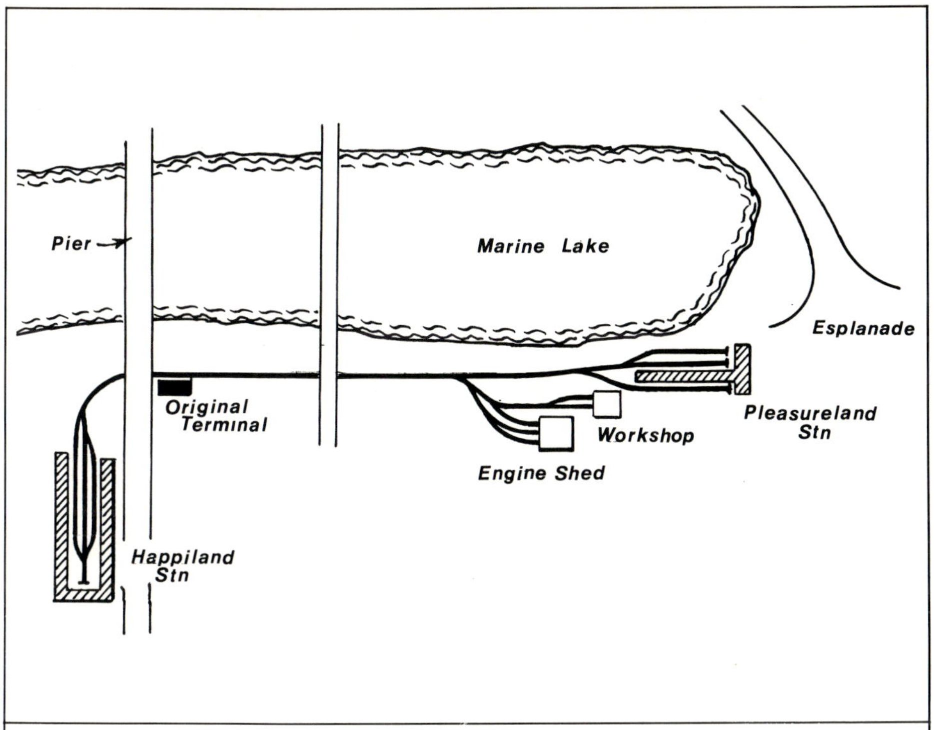

Left:
Lakeside Miniature Railway, Southport: Layout today.

Edward of Wales was sold to Margate, was built to a scale far greater than the one-quarter-size engines formerly seen on 15in-gauge tracks in Britain. It stood nearly 4ft high, measured 17ft 6in over buffers and had a 41-tube boiler fired by coke that developed a working pressure of 120lb/sq in. Stephenson link motion was fitted, together with a notching-up lever giving three forward positions and one reverse. Slide valves operated above the cylinders which had a bore of 4¼in and a stroke of 7in.

Mr Barnes rebuilt the engine shed to house four locomotives and fitted an air compressor to help in raising steam. In 1922 a second Barnes locomotive, No 102 *John*, was introduced to take the place of *George the Fifth*, which was sold to a private owner in Skegness. In 1924 No 103 *Billie* was built and ran on a couple of private lines before being sold to Belle Vue Zoo in Manchester where a new miniature railway had been built. Two years later No 105 *Michael* joined the locomotive list at Rhyl and in 1934 No 106 *Billy*. A further engine had been built for the Margate line in 1928.

Year after year the Rhyl Miniature Railway carried thousands of passengers with two trains in service on the single continuous track each summer. One locomotive was taken out of service for overhaul each year, its boiler being subjected to water pressure tests of 220lb sq in.

In 1969, however, the line closed and the rails were lifted. But eight later years later they were back on new ballast with part of the track realigned along the lake shore. What is more, limited steam was in operation, Alan Keefe Ltd having obtained a 5-year lease to use *Michael* and a diesel for passenger services. In 1980 the locomotive *Billy* was auctioned at Sotheby's in London and sold for £9,000 to Rhyl Town Council. After being stored in the Town Hall it is now on display (inside a glass case) at Rhyl station.

Meanwhile, the line's new owners, local businessmen Leslie Hughes

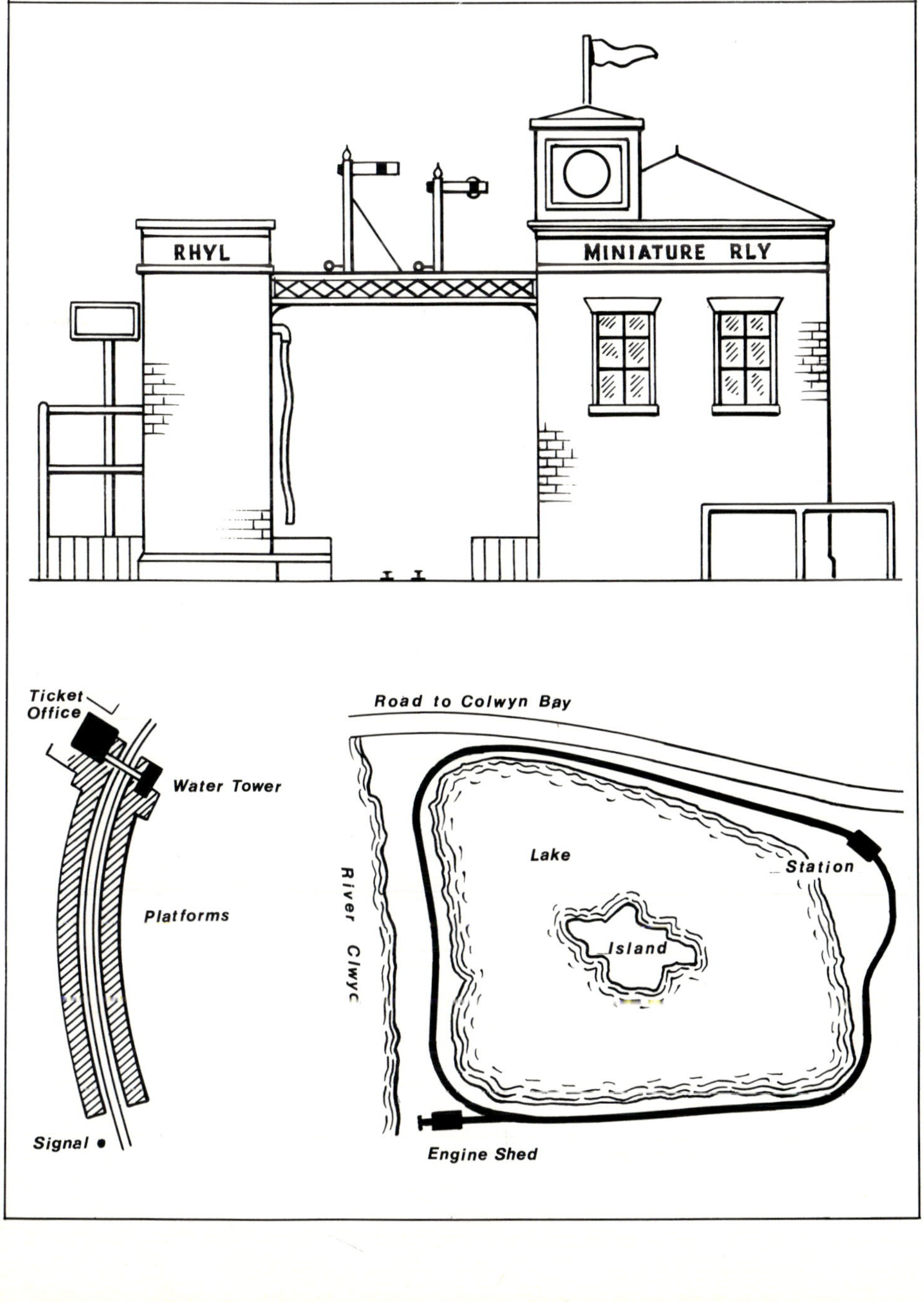

Left:
Rhyl Miniature Railway: Plans and elevation of the 1911 railway.

and Eddie Roberts, brought *Billie* back from Manchester when the zoo closed and renamed it *Louise*. In 1986 all four remaining engines were auctioned at Chester but failed to reach their reserve price of £20,000 each. A legal dispute between the owners, a local ratepayers' group and Rhuddlan District Council, which owns the Marine Lake, complicated matters further.

Joan, however, was back in steam on the Rhyl line in 1987, assisted by a superbly-designed diesel-engined saddle tank named *Clara*. In 1988 and 1989 the maroon-painted *Michael*, now over 60 years old, was in service pulling four of the original coaches during July and August, the only months that running powers are allowed by the council.

Stoking *Michael's* tiny firebox with best Welsh anthracite, has been Ken Dove, son of a renowned Nottingham miniature locomotive builder, who has cared for the railway with dedication and considerable skill for the past 10 years.

'Steam engines are part of Rhyl's heritage', he says. 'Old drivers visit the line just to smell the steam and smoke.'

It is clear that steam at both Rhyl and Southport is on the way back. It would be regrettable if the trains on these historic little seaside lines were to be hauled by diesel locomotives indefinitely.

Steam Locomotive History of Southport and Rhyl 15in-Gauge Miniature Railways

Red Dragon
Year of manufacture: 1909. Builder: Bassett-Lowke. Works No 15. 'Little Giant Class 10'. Type: 4-4-2.

1909: White City Railway.
1911: Rhyl Miniature Railway (renamed *Prince Edward of Wales*).
1920: Dreamland Park Railway, Margate.
1968: In private hands. Later cannibalised.
1987: Parts purchased for Lakeside Miniature Railway, Southport.

King George V
Year of manufacture: 1911. Builder: Bassett-Lowke. Works No 18. 'Little Giant Class 10'. Type: 4-4-2.

1911: Llewellyn's Miniature Railway, Southport.
1913: Rhyl Miniature Railway.
1922: Skegness Miniature Railway.
1930: Southend Miniature Railway.
1938: Belle Vue Zoo Miniature Railway, Manchester.
1955: In store.
1964: Restored by Robin Butterell and John Milner.
1978: In steam at Longleat.
c1980: Steamtown, Carnforth.

Prince Edward of Wales
Year of manufacture: 1912. Builder: Bassett-Lowke. Works No 21. 'Improved Little Giant Class 20'. Type: 4-4-2.

1912: Llewellyn's Miniature Railway, Southport.
1931: Damaged by fire.
1934: Rebuilt and renamed *King George*.
1969: Whorlton Lido Railway, near Barnard Castle.
1988: Lightwater Valley Railway, Yorkshire.

Prince Edward of Wales
Year of manufacture: 1915. Builder: Bassett-Lowke. Works No 22. 'Improved Little Giant Class 20'. Type: 4-4-2.

1915: Fairbourne Railway, Wales.
1922: Llewellyn's Miniature Railway, Southport.
1931: Damaged by fire.
1934: Rebuilt and renamed *Princess Elizabeth*.
1969: In private hands.
1971: Steamtown, Carnforth.

Katie
Year of manufacture: c1896. Builder: Heywood. Type 0-6-0T.

c1896: Eaton Hall Railway, near Chester.
1916: Ravenglass and Eskdale Railway.
1919: Llewellyn's Miniature Railway, Southport.
1922: Fairbourne Railway.

Mighty Atom
Year of manufacture: 1908. Builder: Bassett-Lowke. Works No 11. 'Little Giant Class 10'. Type: 4-4-2.

1908: Sutton Park Railway, Birmingham.
1909: Nancy Exposition, France (renamed *Ville de Nancy*).
1920: Llewellyn's Miniature Railway, Southport (renamed *Prince of Wales*).
1929: Great Yarmouth Miniature Railway.
1937: Sutton Park Railway, Birmingham.
1953: In store.

Joan
Year of manufacture: 1920. Builder: Barnes. Works No 101. Type: 4-4-2.

1920: Rhyl Miniature Railway.
1969: In store.
1970: Belle Vue Zoo Railway, Manchester.
1978: Blenheim Park Railway, Oxfordshire. Also Newcastle-upon-Tyne Town Moor portable line and Steamtown, Carnforth.
1980: Rhyl Miniature Railway.

John
Year of manufacture: 1922. Builder: Barnes. Works No 102. Type: 4-4-2.

1922: Rhyl Miniature Railway.
1978: Alton Towers, Staffordshire (static display).
1985: Whorlton Lido Railway.

Billie
Year of manufacture: 1924. Builder: Barnes. Works No 103. Type: 4-4-2.

1924: Rhyl Miniature Railway.
c1925: Woodland Park Railway, near Rochester (renamed *Michael*).
c1926: Saracens Head Hotel, Warburton, Cheshire (renamed *George V Jr*).
1929: Belle Vue Zoo Railway, Manchester (renamed *Railway Queen*).
1980: Rhyl Miniature Railway (renamed *Louise*).

Billie
Year of manfacture: c1925. Builder: Barnes. Works No 104. Type: 4-4-2.

c1925: Dreamland Park Railway, Margate.
1980: Rhyl Miniature Railway.
1982: Dudley Zoo Railway.
1984. Rhyl (in store).

Michael
Year of manufacture: 1926. Builder: Barnes. Works No 105. Type: 4-4-2.

1926: Rhyl Miniature Railway.
1981: Dudley Zoo Railway.
1982: Rhyl Miniature Railway.

Billy
Year of manufacture: 1934. Builder: Barnes. Works No 106. Type: 4-4-2.

1934: Rhyl Miniature Railway.
1978: Rhyl Town Council (in store).
1986: On display at Rhyl BR station.

Photo Feature:

The East Coast main line in the 1960s

Colin Boocock

We have already commented on the variety of colours in the 1990s on British Rail. Perhaps this compensates for the reduction in variety of trains we can see in any one place on BR nowadays.

It was not always so. In the early 1960s, during three short days of railway photography at the Home Counties end of the East Coast main line, my camera clicked on nine different classes of steam locomotive and eight types of diesels. This selection of prints only skims the surface.

Left:
Peppercorn Class A1 4-6-2 No 60125 *Scottish Union* approaches Potter's Bar with the up 'Tees-Thames', 7.5am from Saltburn to King's Cross on 15 April 1960. *Colin Boocock*

Right:
On the same day *Guillemot*, the 'A4' Pacific No 60020, leans to the curve at Hatfield with the 11am from King's Cross to Edinburgh. *Colin Boocock*

Left:
This fully vacuum braked down express freight photographed near Potter's Bar in April 1960 has 10 Conflat wagons loaded with cement containers at its head, followed by an assortment of four-wheeled vans. The engine is Gresley 'V2' 2-6-2 No 60906. *Colin Boocock*

Below left:
A very grubby 'B1' 4-6-0, No 61363, leaves Potter's Bar with a down King's Cross outer suburban train composed entirely of BR standard non-corridor stock. *Colin Boocock*

Above right:
The original English Electric 'Baby Deltic' Type 2 Bo-Bo diesel electric No D5900 tops the summit at Potter's Bar on 15 April 1960 with an up Cambridge buffet car express. Note the Gresley catering car in the middle of this otherwise BR standard Mk 1 formation. *Colin Boocock*

Centre right:
An unfitted freight (note the grey-coloured wagons leading) gallops along the down local line near Potter's Bar on 21 April 1962. The locomotive is WD 2-8-0 No 90000. *Colin Boocock*

Bottom right:
One of the longest workings for Southern Region locomotives was the return trip on the Cliffe to Uddingston cement trains. A pair of SR Type 3s would take the train from Cliffe in north Kent to join the East Coast main line, where the pilot engine would uncouple. The train locomotive would then take the fully-laden train of around 2,000 tons unaided as far as York. In this photograph taken near Welwyn Garden City on 13 April 1963, No D6575 is returning south from York with the empties. *Colin Boocock*

4A

Photo Quiz

How good is your railway general knowledge? Try to answer all six questions about the photographs on this page.

1. A 'Wessex Electric' Class 442 EMU set — but where is it, and why?
2. A Southern six-car set at Clapham Junction. What exactly is it?
3. This Bulleid 'Battle of Britain' class 4-6-2 looks a little odd. Why?
4. This somersault signal is located at a popular seaside resort. Where?
5. There are only three London Underground stations like this, with a narrow centre island platform flanked by the up and down tracks. Where are they?
6. This former standard five-plank open wagon has lain in this position for many years after being deliberately run off the railway to avoid a potentially worse accident. Can you imagine where it is?

1

2

3

4

5

6

SPOT THE DIFFERENCE

These two photographs are actually not identical. There are 10 differences between them. Can you spot them? Answers are on page 80.

Also, can you say what the locomotive is and where and when the photograph was taken?

N	G	N	I	H	S	A	L	F	C
T	O	R	C	L	A	I	N	I	F
I	R	Y	E	L	L	O	W	S	R
U	I	A	T	V	I	Y	N	H	E
C	A	L	C	T	E	E	E	P	G
R	H	A	C	K	L	L	T	L	A
I	C	N	A	M	W	A	S	A	N
C	U	G	X	O	B	N	O	T	T
J	O	I	N	T	A	M	P	E	R
I	N	S	U	L	A	T	E	D	Y

Wordsearch:

The letter square contains the names of 15 pieces of signalling or track components or associated equipment. They may be written either way, sideways or up and down, or diagonally. Some letters are used more than once. Some names which consist of two words have the words in different places (and directions!) in the square.

See if you can find the 15 names without looking at the key on the answer page (expert grade). Finish it off by looking for the words given on the answer page (standard grade!).

When you have completed the square, if you have done it correctly the remaining letters will form an anagram of a constituent of many bridges and railway sleepers.

The madly fastidious may also find 18 sets of initials of railways, past and present at home and overseas. Can you list them, and say what railways they represent?

Answers are all on page 80.

I THINK IT'S GOT ATHLETE'S FOOT

Answers to Quiz

Photo Quiz

1. The Class 442 unit is at Derby, having been to the carriage works to have its buffet car remodelled.
2. This is the Southern Region's experimental six-car push-pull set made out of coaches from a number of Maunsell express electric multiple-units, mounted on trailer bogies, and was called the '6TC'. It was used to develop the equipment to be fitted to SR locomotives and TC units for the Bournemouth electrification in 1967.
3. No 34049 was experimentally fitted with a modified smoke-deflection arrangement in 1959.
4. This is one of the somersault signals provided by the Belfast & Northern Counties Railway for the main line and branches between Belfast and Londonderry. It stands at Portrush station on Northern Ireland Railways.
5. The three Underground narrow island platforms are at Clapham Common, Clapham North and Angel on the Northern Line.
6. This is the wagon trap near the bottom of the Cromford incline on the former Cromford & High Peak Railway. The trap was there to avoid the effect of runaway wagons on the rope-worked incline running loose into the sidings and buildings at the canal wharf at Cromford. The trackbed is now a heritage trail, and well worth a visit.

Spot the Difference

The locomotive is Talyllyn Railway 0-4-2ST No 1 *Talyllyn*, stored at Towyn Pendre in 1955, awaiting its eventual overhaul.

Wordsearch

The 15 names with signalling and track relevance are:

chair
finial
fish plate
flashing yellow*
gantry
insulated joint*
junction
key
lens
lever
post
signal box*
snow man*
tamper
track circuit*

* These names are split, each word being in a different location and direction in the square.

The remaining letters spell an anagram of CONCRETE.

The initials of railways are:
BN: Burlington Northern
CFL: *Chemins de Fer Luxembourgeois* (Luxembourg Railways)
CFR: *Caile Ferate Romane* (Romanian Railways)
CP: Canadian Pacific
CR: Caledonian Railway
DR: *Deutsche Reichsbahn* (German State Railway — East Germany)
FR: Furness Railway
FS: *Ferrovie dello Stato* (Italian State Railways)
GCR: Great Central Railway
GER: Great Eastern Railway
HR: Highland Railway
IE: *Iarnrod Eireann* (Irish Rail)
IR: Irish Rail
NS: *Nederlandse Spoorwegen* (Netherlands Railways)
RR: Rhymney Railway
SR: Southern Railway
TR: Talyllyn Railway
WSR: West Somerset Railway